Water Distribution in Ancient Rome

Water Distribution in Ancient Rome

The Evidence of Frontinus

Harry B. Evans

Ann Arbor

THE UNIVERSITY OF MICHIGAN PRESS

for Emmy and Larry

First paperback edition 1997

Published in the United States of America by
The University of Michigan Press
Manufactured in the United States of America

2000 1999 1998 1997 4 3 2 1

A CIP catalogue record for this book is available from the British Library.

Library of Congress Cataloging-in-Publication Data

Evans, Harry B.
Water distribution in ancient Rome : the evidence of Frontinus / Harry B. Evans.
p. cm.
Includes bibliographical references and index.
ISBN 0-472-10464-0 (alk. paper)
1. Water—Italy—Rome—Distribution—History. 2. Aqueducts—Italy—Rome—History. 3. Water—Italy—Rome—Distribution—Early works to 1800. 4. Aqueducts—Italy—Rome—History—Early works to 1800. 5. Frontinus, Sextus Julius. De aquis urbis Romae.
I. Title.
TD280.R6E94 1993
628.1'0937—dc20 93-33761
CIP

ISBN 0-472-08446-1 (pbk : alk. paper)

This book has been supported by a grant
from the National Endowment for the Humanities.

Preface to the Paperback Edition

Water in Rome's aqueducts was constantly flowing, and the same can be said for aqueduct studies; much new has been done in this area since the manuscript for the first edition of this book was finished in 1993. The Press has kindly offered me the opportunity to take note of recent work and to correct errors in the earlier version, and, like any author, I want to take full advantage of it.

I have learned much from reviews of the book that have appeared to date, particularly those of John W. Humphrey in *BMCR* 6.5 (1995): 398–402; Christer Bruun in *Phoenix* 49.3 (1995): 279–82; and E.J. Owens in *CR* 41.1 (1996): 146–47 [Shorter notices have also appeared in *Choice* 32.4 (December, 1994): 32–3200 (R.I. Frank) and *JRS* 85 (1995): 312 (Andrew Wilson)]. It is reassuring that all the reviewers make positive comments, but, as expected, the reviews are not without criticism: that the focus on Frontinus' data is too limited (Bruun, 280–82); that the maps in figures 1 and 2 are inadequate (Humphrey, 402; Wilson, 312); that more thorough analysis of the statistics, region by region, is needed (Humphrey, 401–2); that the book ignores an important group of "hard facts" in the names appearing on 280 Roman *fistulae* (Bruun, 281); and that it assumes too strong a connection between urban growth and the development of Rome's water supply (Owens, 146; Bruun, 280–91). Bruun's final comment on the study (282) is probably the most critical of all: "This book raises more questions than it answers. . . ."

Perhaps so, but let me address these points. First, the focus of the book, as its title indicates, is Frontinus' data, not a comprehensive examination of the archaeological and epigraphical evidence for water distribution in ancient Rome; any such study would have resulted in a far longer and far different book. Such an investigation is, of course, needed, but the energetic scholar who undertakes it will also find more questions than answers, given the limitations of the evidence. More important, any such study will not be able to ignore the evidence of

Frontinus' *De aquaeductu,* and that is what this book seeks to demonstrate. For this reason I did not tackle the epigraphical evidence of Roman pipes, much less the problem of the statistics presented by the fourth-century Regionary Catalogues, or the even thornier issue of where their boundaries ran (one of the great battlegrounds of Roman topography). All are subjects worthy of investigation, but that was not the purpose of this study, nor, I think, do the boundary lines of individual regions significantly affect the conclusions drawn from Frontinus here. For the same reason too the maps in figures 1 and 2 are general; they are not intended to be a substitute for more detailed cartographic coverage of the ancient city that readers interested in the topographical issues will certainly want to consult.

Overstressing the connection between demographic factors and the development of the city's water system is a more important issue. Owens acknowledges my statement (p. 135) that aqueducts could be introduced for a number of reasons, not all of them directly related to population growth, but argues that my presentation places too much emphasis on this connection (146–47), a point also made by Bruun; Owens rightly cites Augustus' *naumachia* and the *delectationes* mentioned by Pliny and Vitruvius and supply of private *balnea* (147) as other factors in aqueduct construction. Certainly I could have paid more attention to such evidence, and were I to rewrite the book from scratch, I would present much of the material in a different fashion, giving greater emphasis to such "quality of life" issues. However, the overall thesis still holds: aqueducts up to the time of Frontinus are best linked to Rome's urban growth, and we therefore have much to learn from the data in his text.

Bruun and I disagree fundamentally (although, I should add, quite cordially) on the value of Frontinus' statistics. For this reason he devotes most of his review to comments on methodology, particularly points on which he finds my argument "wanting" (280). I have already cited his remarks about the boundaries of the Augustan *regiones;* let me address other criticism. On the terminus of the Aqua Appia Bruun states that I follow Lorenzo Quilici [*Archeo* 53 (1989): 64] in arguing that the Appia was introduced to provide water to the republican port (281); in fact I had made the point in an earlier article [*AJA* 86 (1983): 403] and expanded upon it in a paper presented at the AIA-APA Annual Meeting in 1987, later published in Trevor Hodge's collection *Future Currents in Aqueduct Studies* (1991). Chronology aside, I still think my argument holds, nor do I agree with Bruun that Filippo Coarelli's suggestion that Appius' line is rather to be associated with the nearby Ara Maxima of Hercules gives a more cogent explanation for its introduction. Bruun also discounts the argument presented here [actually that of M. Gwyn Morgan in *Philologus* 122 (1978)] that senatorial opposition to the extension of the Aqua Marcia to the Capitol in 144

B.C. was prompted by settlement on the hill, suggesting instead that the reason should be sought on religious or political grounds. This is an interesting suggestion, which might lead to fruitful conclusions, but given the paucity of our evidence, Morgan's explanation still seems to me the most convincing.

If this book does raise questions, as Bruun observes, it has also inspired further study of the evidence I would like to cite. In a well-documented paper on the course of the aqueducts within ancient Rome, Rabun Taylor, "*A Citeriore Ripa Aquae:* Aqueduct Crossings in the Ancient City of Rome" [*BSR* 63 (1995): 75–103], cites as "unlikely" (78–79) my argument that at least four aqueducts crossed to the Transtiber via the Pons Aemilius: the Aqua Appia (p. 71), the Anio Vetus and Marcia (p. 107), as well as the Claudia (pp. 121–22). On the Anio Vetus and Marcia, I cheerfully acknowledge that Taylor is correct: Frontinus (*Aq.* 80.2, 81.2) indicates clearly that prior to his reforms those lines did not distribute to Region XI, the location of the eastern bridgehead of the Pons Aemilius; if indeed the Anio Vetus and Marcia distributed to the Transtiber, as Frontinus states, they must have been carried by to the Pons Agrippae further upriver, and their extension to the Transtiber is therefore to be connected with Agrippa's reworking of the water system. Concerning the other two lines, however, my argument still stands: the Appia and Claudia must have been carried by the Pons Aemilius, not to mention the Anio Novus, which also distributed to the Transtiber. Taylor expresses skepticism about multiple lines being carried on a single bridge, but such an arrangement is no more unusual or difficult to imagine than that of the Marcia, Tepula, and Julia brought above ground for over six miles into the city on a single arcade.

The principal thesis in Taylor's article concerns the post-Frontinian Aqua Traiana of A.D. 109 that entered the city at the top of the Janiculum; he argues that Trajan's aqueduct crossed the Tiber on a specially constructed aqueduct bridge south of the Aventine, skirted that hill to the south, crossing the saddle between the Aventine and Little Aventine, eventually reaching Trajan's Baths on the Oppian, and that a section of its arcade is represented on Bufalini's map of Rome published in 1551 (92–102, with a map of its route on 97). How the Aqua Traiana crossed the Tiber remains a mystery, but this reconstruction is, I think, highly fanciful. Even more improbable is Taylor's suggestion (to be sure, made more tentatively) based on *CIL* 6.3865, a fragmentary inscription not found in situ, that the Anio Novus crossed the river via the Tiber Island bridges (82, 91). This paper is, however, a careful piece of scholarship and but one more indication that the last word has not been spoken in making sense of aqueduct lines and water distribution in the ancient city. If this book has given any impetus to such investigations, it has certainly served its purpose.

It remains to cite some new bibliography and bring one notice up to date.

The mill-run on the via Giacomo Medici on the Janiculum (cited on p. 130) has now been published by Malcolm Bell III, "Mulini ad acqua sul Gianicolo," *Archeologia Laziale* 11 (1993): 65–72. In addition, the secondary *castellum* of the Aqua Marcia in the Piazza dei Cinquecento (described as "dismantled" on p. 85 n. 14) is again quite visible, although now inaccesible behind a tall metal fence in a garden area to the right of Stazione Termini. Students of the aqueducts who want to visit the remains described here should definitely take along Peter J. Aicher's new *Guide to the Aqueducts of Ancient Rome* (Wauconda, Ill., 1995), which provides full descriptions, as well as up-to-date maps and detailed directions on reaching the sites. And more work on aqueducts, both in and outside Rome, continues apace; see R.J.A. Wilson, *Tot aquarum tam multis necessariis molibus* . . . Recent studies on aqueducts and water supply," *JRA* 9 (1996): 5–29, for an exhaustive review of recent scholarship. No doubt there are more things to be corrected in the pages that follow, but this preface is already too long. I am very grateful to the Press for reissuing this book and look forward to future controversy from it, as well as better understanding of the problem of water distribution it addresses.

Bronx, New York
Epiphany 1997

Acknowledgments

No one can write a book like this without incurring a great number of debts, and I would like to acknowledge them here. First I am grateful to Fred S. Kleiner, editor in chief of the *American Journal of Archaeology*, for kindly giving me permission to incorporate in chapters 6, 7, 8, and 10 material published in two earlier articles, "Agrippa's Water Plan," *AJA* 86 (1982): 192–204, and "Nero's *Arcus Caelimontani*," *AJA* 87 (1983): 393–99, and to Francis Cairns (Publications) Ltd. for similarly kind permission to incorporate in chapters 3 and 11 material in my paper "Water Distribution in Ancient Rome: *quorsum et cui bono?*" which appeared in A. Trevor Hodge, ed., *Future Currents in Aqueduct Studies* (Leeds, 1991).

Frontinus describes his predecessors in the office of water commissioner as *principes civitatis* (*Aq*. 1), and the same can be said for those scholars who have worked or are working on aqueduct studies; my debts to them are legion. Anyone interested in aqueducts must first acknowledge huge debts to the work of giants like Rodolfo Lanciani, Thomas Ashby, and Esther Van Deman, but I also owe much to contemporaries in the field. Robert H. Rodgers is a singular example here; he very generously made available to me the manuscript of his forthcoming text of and commentary on Frontinus' *De aquaeductu*, soon to be published by Cambridge University Press, thus making it possible for me to incorporate in this study the many contributions he has made to our understanding of an extremely difficult text. I have learned an extraordinary amount from conversations and correspondence with him over the past two years.

Italian scholars currently working on aqueducts have also assisted me greatly. I am most grateful to Giovanna Tedeschi Grisanti and Doriana Cattalini for taking the time to show me the "Trophies of Marius" during their work on its reconstruction, and particularly to Giovanna Tedeschi Grisanti for making available to me page proofs on her study of aqueduct levels at the "Trophies of Marius" and Porta Maggiore, "Primo contributo ad una livellazione urbana sistematica," in *Atti del convegno "Gli antichi acquedotti di Roma."* I also learned quite a few things from Elvira Leone while visiting excavations in progress at the Meta Sudans and from Claudio Mocchegiano Carpano in a visit to the "Sette Sale." I have also gained much from conversations and correspondence with A. Trevor Hodge, Robert B. Lloyd, Christer Bruun, and Peter Aicher. Finally, I am grateful to Alexis Papadopoulos and Philippe Forét for their cartographic assistance in preparing the maps, to Karen Bull-Simonsen Einaudi and the Fototeca Unione at the American Academy in Rome for help in obtaining photographs, and to the Office of Research at Fordham University for its support.

My principal debts, however, are to the two scholars and friends to whom this study is affectionately dedicated. Emeline Hill Richardson introduced me to the study of Roman topography when I was a third-year graduate student at the University of North Carolina-Chapel Hill. Lawrence Richardson, jr, first encouraged me to apply for a Rome Prize Fellowship at the American Academy in Rome, where my year's study in 1972–73 and many subsequent returns changed my life and research interests. Larry's 1977 NEH Seminar for College Teachers at the American Academy, "Problems in the Topography of Ancient Rome," introduced me to the problems of water distribution in the ancient city, a topic with which I have dabbled off and on for the past fifteen years. He has also read more than one version of this study at various stages and has saved me from many errors of judgment and fact. Indeed without the Richardsons this book, such as it is, would have never become a reality. I am deeply grateful to them and others who have assisted me, *principes civitatis* all. The mistakes here, and there will be many, are entirely my own responsibility.

Contents

Abbreviations in the Notes

Abbreviations in the Notes and *Bibliography* follow closely those given in the *Oxford Classical Dictionary,* 2d ed. (Oxford, 1970), ix–xxii, and "Editorial Policy, Notes for Contributors, and Abbreviations," *American Journal of Archaeology* 95 (1991): 1–16. After the first citation of an article or book in the Notes, all following references will be by author's name alone or by name and short title. All short titles can be readily identified in the *Bibliography*. Several works will be cited by special abbreviation as follows:

Ashby	T. Ashby, *The Aqueducts of Ancient Rome* (Oxford, 1935).
Bennett	C.E. Bennett, ed. and trans., *Frontinus, The Stratagems and the Aqueducts of Rome* (Cambridge, Mass., 1925).
Bruun	C. Bruun, *The Water Supply of Ancient Rome: A Study in Roman Imperial Administration,* Commentationes Humanarum Litterarum 93 (Helsinki, 1991).
CAR	*Carta Archeologica di Roma* (Florence, 1962–).
CIL	*Corpus Inscriptionum Latinarum*.
Currents	A.T. Hodge, ed., *Future Currents in Aqueduct Studies* (Leeds, 1991).
Grimal	P. Grimal, ed. and trans., *Frontin, les aqueducs de la ville de Rom* (Paris, 2d ed., 1961).
FUM	E. Rodriguez Almeida, *Forma urbis marmorea: aggiornamento generale* (Rome, 1980).
FUR	G. Carettoni, A.M. Colini, L. Cozza, and G. Gatti, *La pianta marmorea di Roma antica (Forma Urbis Romae),* 2 vols. (Rome, 1960).

Hodge	A.T. Hodge, *Roman Aqueducts and Water Supply* (London, 1992).
Jordan	H. Jordan, *Topographie der Stadt Rom im Alterthum*, 2 vols. in 4 (Berlin, 1871–85; repr. Rome, 1970).
Kunderewicz	C. Kunderewicz, ed., *Sex. Iulii Frontini, De Aquaeductu Urbis Romae* (Leipzig, 1973).
Lanciani	R. Lanciani, *I commentarii di Frontino intorno le acque e gli acquedotti, MemLinc* ser. 3.4 (1881; repr. Rome, 1975).
LFUR	R. Lanciani, *Forma Urbis Romae* (Milan, 1893–1901; repr. Rome, 1990).
Lugli	G. Lugli, *I monumenti antichi di Roma e suburbio*, 3 vols. (Rome, 1930–38).
Nash	E. Nash, *Pictorial Dictionary of Ancient Rome*, 2d ed. (New York, 1967).
Pace	P. Pace, *Gli acquedotti di Roma e il "De Aquaeductu" di Frontino. Contesto critico, versione, e commento* (Rome, 1983).
Platner-Ashby	S.B. Platner and T. Ashby, *A Topographical Dictionary of Ancient Rome* (Oxford, 1929; repr. London, 1950; Rome, 1965).
RE	A. Pauly, G. Wissowa, and W. Kroll, eds., *Real-Encyclopädie der classischen Altertumswissenschaft.*
Richardson	L. Richardson, jr, *A New Topographical Dictionary of Ancient Rome* (Baltimore and London, 1992).
Roma sotterranea	R. Luciani, ed., *Roma sotterranea*, Exhibition Catalog (Rome, 1984).
Trionfo	G. Pisani Sartorio and A. Liberati Silverio, eds., *Il trionfo dell' acqua: acque e acquedotti a Roma, IV sec. a.C.–XX sec.*, Exhibition Catalog (Rome, 1986).
Van Deman	E.B. Van Deman, *The Building of the Roman Aqueducts* (Washington, D.C., 1934).
Wasserversorgung	Frontinus-Gesellschaft e.V., *Wasserversorgung im antiken Rom* (Munich, 1982).

Introduction

One of the long-standing problems in Roman topography is that of water distribution in the ancient city, specifically how water brought into Rome by its aqueducts was delivered to its neighborhoods and districts. Rome's aqueducts have long fascinated visitors to the city, and its water system has been the subject of intense scholarly interest since the publication of Rodolfo Lanciani's monumental commentary on Frontinus' treatise *De aquaeductu urbis Romae,* a study that has been aptly described as "the starting point of modern topographical work."[1] Scholars such as Thomas Ashby and Esther Van Deman have produced magisterial studies on the courses of the aqueducts outside the city,[2] and the imperial administration of Rome's water supply has most recently been extensively examined by Christer Bruun.[3]

In addition, aqueduct systems throughout the Roman world are now receiving fresh attention through publications of the Frontinus-Gesellschaft in Germany[4] and the ongoing work of scholars in Italy, North America, and

---

1. Ashby, 9 (see Abbreviations).

2. Ashby; Van Deman. The contributions of these two scholars will never be superseded, because much of the physical evidence documented by them has now been lost as a result of Rome's rapid expansion into the countryside following World War II.

3. Bruun, 393–419, also provides a full bibliography of work to date. Cf. also O.F. Robinson, "The Water Supply of Ancient Rome," *Studia et Documenta Historiae e Iuris* 46 (1980): 61–78; id., *Ancient Rome: City Planning and Administration* (London, 1992) 95–105; and C. D'Amato, "L'Amministrazione delle acque in età romana," in *Trionfo,* 176–87.

4. Three volumes have now appeared in its "Geschichte der Wasserversorgung": *Die Wasserversorgung im antiken Rom* (Munich, 1982) includes a German translation of Frontinus, photographs, and plans; *Die Wasserversorgung antiker Städte: Geschichte der Wasserversorgung* vols. 2 and 3 (Mainz, 1987–88) focus on ancient Pergamum, the structural elements of water systems, their administration, and fountains, as well as

elsewhere.[5] Very little, however, has been done to determine what happened to Rome's water after it reached the city. Its final distribution remains essentially an unsolved problem, a fact observed by Giuseppe Lugli more than fifty years ago,[6] indeed one that is even now regarded as insoluble.[7]

Our ignorance about water delivery in ancient Rome is in many ways not surprising, given the state of the evidence available. Frontinus' treatise on Rome's water system, our primary source, survives only in a faulty manuscript tradition, and it furnishes statistics that are at times conflicting on the volume and delivery of individual aqueduct lines. As a text of crucial importance to ancient historians and students of technology, *De aquaeductu urbis Romae* has attracted new scholarly attention in the last twenty years; the most recent Teubner edition, that of Cezary Kunderewicz published in 1973, has received both generally favorable reviews and critical comment.[8] Its appearance has prompted further textual study and scholarly work, activity that no doubt will continue and needs to continue.[9]

To be sure, Frontinus' statistics on water distribution in the ancient city are

---

water systems in northern Europe. See also K. Grewe, *Planung und Trassierung römischer Wasserleitungen* (Wiesbaden, 1985); and R. Tölle-Kastenbein, *Antike Wasserkultur* (Munich, 1990).

5. Hodge presents the most recent comprehensive study of Roman aqueducts and water systems, with a full bibliography, 359–88. Cf. also the comprehensive bibliography in Bruun, 393–419. The 1986 Exhibition Catalog *Il trionfo dell' acqua* contains many articles of value on aqueducts of ancient, Renaissance, and modern Rome. Two recent general studies focusing on the aqueducts of Rome itself have also appeared: that of G. Panimolle, *Gli acquedotti di Roma antica* (Rome, 1963); and that of Pace, which focuses primarily on engineering but also gives a text and Italian translation of Frontinus. L. Quilici, "Gli acquedotti di Roma," *Archeo. attualità del passato* 53 (July, 1989): 50–97, provides a general review of the Roman aqueducts, with photographs and maps. In addition, the 1987 Annual Meeting of the Archaeological Institute of America included a special colloquium on ancient water systems, the papers from which were published in 1991 (*Currents*).

6. G. Lugli, review of the books by Van Deman and Ashby, *Gnomon* 13 (1937): 98.

7. Bruun, 117 and n. 7, describes any attempt to deal exhaustively with the question as "quite futile," "respectfully" doubting Lanciani's comment in 1880 that "con un poco di pazienza" the distribution of the Claudian aqueducts could be traced within the city (Lanciani, 363).

8. C. Kunderewicz, ed., *Sex. Iulii Frontini, De aquaeductu urbis Romae* (Leipzig, 1973). Cf. the reviews by M. Lenoir, *REL* 53 (1975): 453–54; O.A.W. Dilke, *JRS* 67 (1977): 247–48; L. Callebat, *RPhil* 49 (1975): 326–27; and P. Grimal, *Gnomon* 47 (1975): 624–25.

9. R.H. Rodgers is currently preparing a new text and scholarly commentary on the work. Cf. R.H. Rodgers, "The Textual Tradition of Frontinus' *De Aquaeductu Urbis Romae:* Preliminary Remarks," *BICS* 25 (1978): 101–6; id., "Frontinus on Aqueducts: Textual Temptations," *BICS* 30 (1983): 131–36; and Bruun, 381–84.

partial and limited; in addition, because he wrote his treatise in A.D. 97, shortly after he assumed the post of *curator aquarum* or water commissioner under Nerva (*Aq*. 1–2, 102.17, hereafter cited without title), Frontinus gives no information at all on the two aqueducts introduced at later dates, the Aqua Traiana of A.D. 109 and Aqua Alexandriana of the early third century.

Moreover, although most of Rome's aqueducts can be traced to terminal reservoirs within the city, there is almost a total lack of archaeological evidence as to what happened to their water after that point. Most of the lead pipes used to convey it no longer survive, having been stolen and used for other purposes in medieval and modern Rome,[10] and excavation reports have been rightly described as old and "generally quite casual," making it difficult to draw convincing conclusions about watercourses within the city.[11] In addition, the full extent of terracotta pipes and free-flow channels in Rome remains largely unknown.[12]

The problem is still worth attention, because much more can be done with the evidence we have: Frontinus and other literary sources; archaeological remains, however meager; and our understanding of ancient Rome, which has been significantly advanced by the work of many archaeologists and topographers during the last fifty years, since the studies of Ashby and Van Deman were published. Recent investigations of specific aqueducts and branch lines have indeed shown that careful examination of all the evidence—literary, archaeological, and topographical—can yield significant results.[13] This study seeks to advance on these contributions.

Our starting point is the evidence of Frontinus, a treatise indispensable for serious study of Rome's water supply. Because this source will be cited throughout, chapter 1 presents a translation of *De aquaeductu urbis Romae* based primarily on R.H. Rodgers' forthcoming critical edition and commentary.[14] Since

---

10. For the loss of these conduits see Lanciani, 414–15, who also included in his commentary a "Silloge Epigrafica Acquaria" (423–502) cataloging 596 pipes found in and around Rome. This work is long superseded by that of H. Dressel, *CIL* 15.2,1, 906–93; and, most recently, by that of Bruun, 207–380.

11. Bruun, 117, points out that only three of nine planned volumes of the *Carta Archeologica di Roma* have been published to date.

12. On free-flow channels see Bruun, 118–23; on open conduits in other distribution systems, see Hodge, 315–17. On terracotta piping, recommended by the Roman architect Vitruvius (*De arch*. 8.6.1–8), see Bruun, 125–39; and Hodge, 113–17.

13. Cf. R.B. Lloyd, "The *Aqua Virgo, Euripus* and *Pons Agrippae*," *AJA* 83 (1979): 193–204; H.B. Evans, "Agrippa's Water Plan," *AJA* 86 (1982): 401–11; id., "Nero's *Arcus Caelimontani*," *AJA* 87 (1983): 392–99; F. Coarelli, "L'*urbs* e il *suburbium*," in A. Giardina, ed., *Società romana e impero tardoantico*, 2 vols. (Bari, 1986), 2:43–45.

14. The only other English version is that of C.E. Bennett's Loeb edition (Cambridge, Mass., 1925), based on the earlier translation of C. Herschel, *The Two Books on the*

the Latin text of Frontinus is not well preserved, any English version of it will not be without problems and disputed passages. I have cited these, when appropriate, in the notes. The translation is included mainly as an aid to the reader, who will need ready access to pertinent sections of *De aquaeductu.*

While this study begins with a complete translation of Frontinus' treatise, it is not intended to be a comprehensive commentary on *De aquaeductu*. Indeed the recent work of Bruun and the contributions of scholars within the Frontinus-Gesellschaft have done much to fill that need, not to mention Rodgers' forthcoming text and commentary. Preliminary remarks are therefore in order concerning aims and methodology.

Although I am primarily concerned with water delivery, I shall not attempt to make sense in modern terms of the volumes that Frontinus reports as *quinariae,* a measurement based on the capacity of the *fistula quinaria,* a pipe one and a quarter Roman digits in diameter (25.1–2). The problem has been much debated, and modern estimates of the total water delivery within the ancient city vary greatly, from more than a million cubic meters a day to five hundred thousand to six hundred thousand cubic meters. Such estimates are, of course, of limited value, since there is no sure way to determine the population of ancient Rome at any given time, and therefore no means to establish an average daily volume per inhabitant.[15]

Several investigations have been made on the capacity and volume delivered by the Roman aqueduct system: a study by Claudio di Fenizio almost eighty years ago focused on investigations of the *quinaria* measurements given by Frontinus,[16] and a second, much more recent, by Deane Blackman based its findings on extensive measurement of channel widths and flow depths of extant water conduits of four major aqueducts.[17] Both studies through independent methodologies have resulted in a "modest agreement" concerning the total volume of water delivered by the four major aqueducts of imperial Rome. Such "agreement" does not resolve the problems connected with determining the

---

Water Supply of the City Rome (Boston, 1899; repr. 1973). There is a French translation in P. Grimal's Budé edition (Paris, 1961), and German versions of the Kunderewicz text can be found in *Wasserversorgung,* 81–120, and M. Hainzmann, *Wasser für Rom: Die Wasserversorgung durch Aquädukte* (Zurich, 1979). Pace, 235–314, gives an Italian version.

15. Bruun, 98–100.

16. C. di Fenizio, "Sulla portata degli antichi acquedotti romani," *Giornale del Genio Civile* 14 (1916): 227–331, followed by Hodge, 299–300; and, with slight modifications, by Pace, 61–70. Bruun, 385–87, gives a summary of di Fenizio's methods and conclusions.

17. D.R. Blackman, "The Volume of Water Delivered by the Four Great Aqueducts of Rome," *PBSR* 46 (1978): 52–72.

volumes of water distributed, much less establish this delivery in modern units of measurement,[18] but it does strengthen the credibility of Frontinus' figures.

The last few years have seen further investigation into Frontinus' treatise and its statistics, specifically whether he could measure the *quinaria* by scientifically accurate methods, and, if so, how such measurements were taken.[19] These studies confirm that while the *quinaria* cannot be converted into modern units of measurement, Frontinus regarded it as an accepted unit internally consistent and applicable to the statistics he reports in *De aquaeductu* (34.2–3). We can therefore use his data to determine the relative capacities and delivery of individual lines within Rome's water system as a whole.[20] Moreover, even if exact deliveries cannot be determined for individual aqueducts, given the faulty state of the MS tradition and possible variations in the size of the *quinaria* from line to line, Frontinus' statistics do reflect administrative practices in effect when he assumed office and therefore tell us much about hydraulic planning for the city.[21]

In any case, we must recognize that the statistics of *De aquaeductu urbis Romae* are not only data supplied by a knowledgeable source (albeit in a sometimes faulty text), but also, and more important, the primary statistics available to us. Lacking others, we must use Frontinus' *quinariae* figures in any attempt to make sense of the distribution of water within the ancient city. Our problem then is how each aqueduct delivered its water and what areas of the city each line served. How much water within the total supply did each aqueduct distribute? What needs did it supply? What role did each line play within the overall hydraulic system of Rome, both at the time of its introduction and at the end of the first century A.D., when Frontinus wrote his treatise?

Frontinus' data indicate that approximately one-third of the total delivery of Rome's aqueducts was distributed outside the city, 4,063 *quinariae* from a total of 14,018 (78.2). This study will not, however, address directly the problem

---

18. R.H. Rodgers, "*Copia Aquarum:* Frontinus' Measurements and the Perspective of Capacity," *TAPA* 116 (1986): 353–60, 357 n. 19, rightly observes that "modest agreement" does not establish overall certainty and that di Fenizio's calculations were presented to give only a minimal value for the *quinaria*.

19. A.T. Hodge, "How did Frontinus Measure the *Quinaria*?" *AJA* 88 (1984): 205–16; Rodgers, "*Copia Aquarum,*" 353–60.

20. Cf. Bruun, 372, in discussing the volume of water delivered in ancient Rome: "Since the Romans themselves were not able to calculate exactly the volume of flowing water, there is not much point in relying on the figures of Frontinus. Moreover, these figures are sometimes contradictory. . . ." Such observations are certainly valid in attempts to determine the actual volume of delivery in modern terms, but that is not the purpose of this study.

21. Bruun, 100 n. 19.

of supplies to the *suburbium*, a subject which merits full examination on its own, except where such deliveries *extra urbem* are relevant to distribution of individual lines within Rome itself.[22] In addition, since Frontinus' *De aquaeductu* is a treatise on the history and organization of Rome's water system, and not a manual on Roman hydraulics,[23] any study of water distribution based primarily on its evidence must be limited in focus. We shall not attempt to investigate the technology of the aqueduct system or to explore problems of Roman hydraulic engineering, such as the operation of aqueduct lines, siphon systems, and distribution tanks, all topics covered in standard introductory works on ancient engineering.[24] However, we must begin with a few observations on Roman water distribution, to explain terms found throughout Frontinus' text and in subsequent chapters.

Water Distribution: The Hydraulics

Aqueducts in the Roman world worked basically on the principle of gravity flow, quite different from the closed pipe systems used today. Water was carried through an enclosed conduit (*specus* or *rivus*), which usually ran underground, from its source to its terminus or distribution tank (*castellum*). Above ground the line could be carried on a raised embankment (*substructio*) or on an arcade or bridge. Along the line could be incorporated any number of settling tanks (*piscinae*) to remove foreign matter or sediment between its source and *castellum*. At points along the course a subsidiary line (*ramus*) could be introduced to supply additional water, in order to augment the capacity of the line. In addition, a subsidiary or branch line (also designated a *ramus*) could begin from a main conduit before it reached its *castellum*.

At its distribution point the water was divided for delivery through pipes (*fistulae*), either tile or lead, which were attached to the *castellum* by a delivery

22. For a brief but informative discussion of suburban distribution southeast of Rome see Coarelli, "*Suburbium*," 43–45. On distribution in Tiburtine territory see Z. Mari, *Tibur, pars quarta, Forma Italiae* 35 (Florence, 1991), 40; and H.B. Evans, "*In Tiburtium usum:* Special Arrangements in the Roman Water System," *AJA* 97 (1993): 447–55.

23. This point, already made by Grimal, xiv; Bruun, 17–18, 369–70; and Hodge, 16–18, will be explored further in chapter 2.

24. Hodge, 67–331, gives a comprehensive treatment of the technology involved; for a summary, cf. Hodge, 1–3. See also R.J. Forbes, *Studies in Ancient Technology*, 9 vols., 2d ed. (Leiden, 1964–72), 1:149–94; G. Garbrecht, "Wasserversorgungstechnik in römischer Zeit," in *Wasserversorgung*, 9–31; J.G. Landels, *Engineering in the Ancient World* (Berkeley and Los Angeles, 1981), 34–57; Robinson, "Water Supply," 44–46; and P. Pace, "Tecniche di conduzione e distribuzione dell' acqua in epoca romana," in *Trionfo*, 138–51.

fitting or nozzle (*calix*), then laid, usually below ground level, along major streets. For our understanding of how an aqueduct *castellum* worked we are indebted in part to a chapter of Vitruvius' treatise on architecture (*De arch.* 8.6.1–2):[25]

> When it [the water] has reached the city, build a reservoir with a distribution tank in three compartments connected with the reservoir to receive the water, and let the reservoir have three pipes, one for each of the connecting tanks, so that when the water runs over from the tanks at the ends, it may run into the one between them. From this central tank, pipes will be laid to all the basins and fountains; from the second tank, to baths, so that they may yield an annual income to the state; and from the third, to private houses, so that water for public use will not run short; for people will be unable to divert it if they have only their own supplies from headquarters. This is the reason why I have made these divisions, and also in order that individuals who take water into their houses may by their taxes help to maintain the conducting of the water by the contractors.

Vitruvius' description of tripartite division for basins and fountains (*lacus et salientes*), baths (*balneae*), and private houses (*privatae domus*) is generally repeated in standard works on Roman hydraulics and is indeed even accepted by many scholars as a canonical arrangement.[26] It does not, however, agree with the analysis of distribution given by Frontinus in *De aquaeductu*, to be discussed below. Such disagreement is not an insurmountable problem, since Frontinus wrote a good 100 years after Vitruvius, and distribution plans within Rome may well have been modified during the first century A.D. Moreover, Vitruvius' treatise, highly theoretical in nature, is frequently in conflict with what we know of actual practice in the Roman world.[27]

The archaeological evidence on Roman *castella* available to us makes the problem even more complicated: remains of *castella* that survive at Pompeii and Nîmes indicate that these tanks distributed their water, not strictly according

---

25. I follow the translation of M.H. Morgan, *Vitruvius, The Ten Books on Architecture* (Cambridge, Mass., 1914).

26. Two examples are M. Hainzmann, *Untersuchungen zur Geschichte und Verwaltung der stadtrömischen Wasserleitungen* (Ph.D. diss., Universität Graz, 1973), 20–21; and G. Tedeschi Grisanti, *I "Trofei" di Mario: il ninfeo dell' Acqua Giulia sull' Esquilino* (Rome, 1977), 36–38; Hodge, 13–16, 280–82, comments critically on such acceptance of Vitruvius.

27. Bruun, 16 n. 16; Hodge, 15–16.

to function or subsequent use, but rather according to the geography of the areas served: pipes from the *castellum* carried water along main streets to designated neighborhoods, and the same branch lines supplied both public basins and private homes.[28] From main reservoirs a number of branch lines were laid which themselves terminated in their own *castella,* whence the water was distributed further.[29] We cannot assume that the relatively simple distribution net in Pompeii paralleled arrangements in Rome itself,[30] but the large number of *castella* that Frontinus records in the city at his time, 247 in all (78.3), suggests that geographical distribution in individual regions or neighborhoods was a standard arrangement followed there.

Water Distribution: Frontinus' Data

Frontinus provides statistics on water distribution (78–86) both outside (*extra*) and inside (*intra*) Rome,[31] before he assumed responsibility for his office (77.1–2). He also gives a division of volumes in *quinariae* for each line according to function or supply: waters were furnished *nomine Caesaris,* to *privati,* and for *usus publici,*[32] a category that he further divides into four areas, *castra, opera*

---

28. For the distribution system at Pompeii see L. Richardson, jr, *Pompeii: An Architectural History* (Baltimore and London, 1988), 51–63; Hodge, 282–84; and H. Eschebach, "Die innerstädtische Gebrauchwasserversorgung, dargestellt am Beispiel Pompejis," in J.-P. Boucher, ed., *Journées d'études sur les aqueducs romains* (Paris, 1983), 79–132; for that of Nîmes cf. A. Grenier, *Manuel d'archéologie gallo-romaine*. Vol. 4.1, *Les monuments des eaux* (Paris, 1959), 97–98; and Hodge, 284–89.

29. Cf. Richardson, *Pompeii,* 55: "the main purpose of the castellum was to direct water to different quarters at different times." Frontinus himself describes such distribution in the case of the Aqua Crabra outside Rome (9.5).

30. Bruun, 124–25; Hodge, 291–93, 300.

31. Bruun, 147–48, notes that Frontinus uses these phrases without specification but argues that the *urbs* in the first century A.D. may well have extended beyond the line of the later Aurelian Wall. He suggests that the seventh milestone, where several aqueducts had their settling tanks (19.1), may have marked for Frontinus the limit between the *urbs* and surrounding *ager*. The distinction itself between city and *suburbium* in antiquity was an ambiguous one; see E. Champlin, "The *Suburbium* of Rome," *AJAH* 7 (1982): 97–98.

32. G. Tedeschi Grisanti, "I terminali degli acquedotti," in *Trionfo,* 151, considers this tripartite division as roughly parallel to that given by Vitruvius (cited above, n. 25), arguing that water distributed *nomine Caesaris* in Frontinus' statistics includes that assigned to *balneae* by Vitruvius a century earlier: by Frontinus' time, she suggests, the emperor would have assumed responsibility for the supply to baths. This is a tidy way of resolving the discrepancy between the two authors, but *balneae* would seem more likely to be included among either *opera publica* or grants to *privati,* not the category of installations supplied *nomine Caesaris*. Frontinus cites the special status of grants to *balneae* (107–8), a provision that would be unnecessary if they were part of distribution *nomine Caesaris,* and also singles out the Euripus Virginis, part of Agrippa's bath

publica, munera, and *lacus.* To make sense of his statistics we must explore the meaning of these terms. Interpretation of them is far from secure, and is further complicated by the failure of Frontinus' categories to correspond to inscriptional evidence from extant *fistulae*. Indeed very few pipe stamps indicate the purpose or destination of the water the *fistulae* carried.[33]

Water distributed *nomine Caesaris* appears to have been that delivered at the emperor's disposition to supply the imperial palace and buildings directly under his control, to include staff headquarters and offices.[34] It is worth noting that Frontinus includes in this category the delivery of 254 *quinariae* to the *naumachia Augusti* in Region XIV (85); distribution *nomine Caesaris* could therefore include delivery to large structures under imperial control, even if constructed for public benefit or entertainment.[35]

That distributed to *privati* went to citizens granted water for private use by favor of the emperor (*beneficio principis,* 3.2, 69.6, 74.4), an arrangement that seems to have been instituted by Marcus Agrippa, who first published *commentarii* fixing distribution to private parties (98.2). According to Frontinus, in the early republic (*apud antiquos*) almost all water had been used for public purposes (94.3). In his time, however, *privati* received water through written authorization (*epistula a principe* 105.1) and were subject to a water tax (118.1–3; cf. Vitruvius 8.6.2, cited above).[36]

Usus publici are best discussed according to the four subcategories listed by Frontinus. *Opera publica* or public works appear to be not only public buildings but all structures or monuments erected for public use, benefit or amusement that were not directly the property of the emperor or his *fiscus:* these include baths, meeting places, theaters and amphitheaters, circuses, markets, and porticoes.[37] Among the bath complexes were, of course, the great imperial *thermae,* but this category may have also included smaller bathing establishments or *balneae,* which existed in great numbers, from the evidence of the fourth century Regionary Catalogs.[38] Although there is very limited

---

complex in the Campus Martius, as one of the *opera publica* supplied by the Aqua Virgo (84.3).

33. Bruun, 256.

34. Cf. Lanciani, 581; Hainzmann, *Untersuchungen,* 28 n. 1; Grimal, 39 n. 1; Robinson, "Water Supply," 78.

35. We might expect the naumachia to be one of the 95 *opera publica* supplied in the city (78.3), but Frontinus' statistics indicate otherwise.

36. Cf. Hainzmann, *Untersuchungen,* 60–61.

37. Lanciani, 581, followed by Grimal, 39 n. 2; Hainzmann, *Untersuchungen,* 28 n. 1; and Bruun, 102 n. 25.

38. A. Nordh, *Libellus de Regionibus Urbis Romae, SkrRom* ser. II, iii (Rome, 1949), 105. The Breviarium of the Regionaries gives a total of 856 *balneae* within the city, a figure that is suspiciously high, perhaps concocted to conform to a predetermined scheme

evidence for supply to *balneae* in extant *fistulae* inscriptions,[39] such establishments certainly received aqueduct water. Frontinus himself indicates that they had once enjoyed special status among water grants (107–8), and Martial (6.42.18) describes the *thermulae* of Claudius Etruscus as an installation providing water of both the Aqua Marcia and Aqua Virgo to its patrons. *Balneae* are also to be included in Frontinus' category of distribution to *privati*, since many of the complexes were privately owned. Frontinus cites only ninety-five *opera publica* in the city at his time (78.3), and parallels of density from Pompeii and Ostia would indicate that several hundred bathing establishments must have existed in imperial Rome.[40] Also worth noting is that Frontinus specifically cites among *opera publica* the Euripus Virginis, a monumental stream of running water constructed by Agrippa in the Campus Martius (84.3).[41]

Castra would normally indicate military camps and installations, but the number Frontinus lists for them in the city, eighteen in all (78.3), and the wide range in volumes of *quinariae* distributed to them suggest that the term should also be applied to paramilitary or even nonmilitary posts or stations with dormitory arrangements, such as barracks of the *vigiles* or fire brigade, installations like the Castra Lecticariorum and Tabellariorum, and offices of various sizes.[42] Recent studies have shown that such installations had no fixed form or size; those of the *vigiles* were sometimes situated in houses or earlier buildings specifically adapted for the functions accommodated.[43] By Frontinus' time *castra* seem to have been considered a distinct type of public building; otherwise we might expect him to list distribution to them under the category *nomine Caesaris*. Certainly individual *castra* varied in size and water needs because of the wide discrepancies in delivery to individual installations: the Aqua Appia supplied four *quinariae* to one (79.2), the Anio Vetus fifty to another (80.2), the Marcia forty-two *quinariae* to four (81.2), and so forth.[44]

---

of two *balneae* and four *lacus* for each *vicus* within the individual regions. On the difficulties of using the Regionaries as a topographical source see Nordh, 58–67; and G. Hermansen, "The Population of Imperial Rome: the Regionaries," *Historia* 27 (1978): 131–45, 165–69.

39. Bruun, 75–76.

40. Bruun, 74 n. 48.

41. Lloyd, 196–97; Hainzmann, *Untersuchungen*, 28 n. 1. The number 95 is an emendation by Poleni adopted by most editors to make sense of the arithmetic in chapters 79–86.

42. For the mysterious Castra Lecticariorum, Tabellariorum, and Victimariorum, cited only in the Breviarium of the Regionary Catalogs (Nordh, 106), see Jordan, 2:71–72.

43. For the *castra* of the *vigiles* see J.S. Rainbird, "The Fire Stations of Imperial Rome," *PBSR* 54 (1986): 147–69.

44. Cf. Bruun, 245–56, who suggests that *castra* may refer to another category of

The interpretation of *munera* is more problematic. In the text they appear to be ornamental fountains giving a display of their water (3.2), a definition accepted by most commentators on the passage.[45] Frontinus' notice that the staff of the *cura aquarum* were stationed at *munera* as well as at *castella* throughout the city (117.3) indicates that they could also serve as a reserve supply of water in case of emergencies.[46] Moreover, common citizens may well have been able to draw water from the *munera*, given the quantity of water supplied to them.[47] The large number of *munera* listed, thirty-nine in all (78.3), and the varying volumes supplied to them suggest a wide range of size; they should not necessarily be thought of as fountain displays on a scale like that of the "Trophies of Marius," the one example usually cited within Rome. *Munera* could also serve a double function as distribution *castella:* recent excavations have confirmed that the "Trophies of Marius" channeled its water into a delivery system after its monumental display.[48]

Finally Frontinus gives distribution figures for *lacus*, public basins located throughout the city, 591 in all (78.3). That the term indicates open basins is clear from another reference (94.3), describing *aqua caduca*, the overflow from *lacus* taxed for use in baths and fullers' establishments (94.3). Closely associated, indeed inseparable, from the *lacus* were the *salientes* or water outlets supplying the basins located underneath the taps (87.5). The text of a senatorial decree of 11 B.C. quoted by Frontinus (104.2–3) also connects the functioning of *publici salientes*, which fed the public basins, with the constant delivery of water in Rome.[49]

Rome's aqueduct system evolved on a piecemeal basis over a period of five hundred years, as new lines were added to meet existing demands. We must

---

"public water post" or "local cistern" and that Frontinus "presumably" included supplies to military camps under the category *nomine Caesaris*. This, however, seems unlikely; we would expect a much greater uniformity of volume if *castra* were cisterns or storage facilities. Moreover, we would also expect Frontinus to gloss this unusual meaning of the word, in much the same way he glosses his technical use of *munera* (3.2). Even more problematic is the use of the term to indicate a "local cistern," if *castra* were supplied through *castella:* we would expect the diminutive form to indicate the smaller reservoir or tank, not the larger one within the distribution net.

45. Lanciani, 581; Grimal, 65–66. Cf. P. Grimal, review of F.W. Shipley, *Agrippa's Building Activities in Rome*, *REL* 12 (1934): 240, followed by Robinson, "Water Supply," 77: *munera* are a euphemism for public latrines.

46. Cf. Grimal, 66 n. 3.

47. Bruun, 102–3.

48. G. Tedeschi Grisanti, "Le mostre degli antichi acquedotti: i 'Trofei di Mario'," in *Trionfo*, 126–35; cf. Lanciani, 581; Grimal, 65–66.

49. Pliny *HN* 36.121, in discussing Agrippan improvements to the water system, appears to distinguish *salientes* from *lacus*. For a discussion of the problem cf. Bruun, 105–6 n. 43; Jordan, 2:61; Lanciani, 581; and Hainzmann, *Untersuchungen*, 28 n. 2.

therefore examine each line separately, according to the chronology of its introduction. Our discussion will focus on what is known about the route of the line within the city, the position of its terminal *castellum,* and the distribution it carried out, both at the time of its introduction and when Frontinus wrote his treatise. We turn first to the text of *De aquaeductu,* and then will review the organization of Frontinus' treatise and the reasons for its publication.

CHAPTER 1

Sextus Julius Frontinus, *On the Water Supply of the City of Rome*

1. Every job assigned by the emperor demands an unusually high degree of concern. As for myself, a task entrusted to me arouses me to give it not only my attention but even my devotion, whether from a natural sense of responsibility or from what is expected of honest diligence. The post with responsibility for the water supply has now been assigned to me by Nerva Augustus, an emperor whom I don't know whether to describe as more attentive or more devoted to the state. This is a task that concerns both the needs and health of the city, and even its safety, one which has been managed always by leading men of the state. For these reasons I judge it first and foremost, as I have made it my practice in other assignments, to get to know what I have undertaken.

2. Indeed I would believe that there is no surer basis for any performance of office. Otherwise it is impossible to decide what should be done and what should be avoided. Moreover, for a respectable individual there is nothing so disgraceful than to carry out a job assigned to him from the instructions of subordinates, something that must happen whenever ignorance of the task forces him to have recourse to the knowledge they have gained from experience. Even though the roles subordinates play are necessary, nonetheless for administration they are employed as the "hands and tools" of the director. (2) For this reason, following my practice in earlier assignments, I have put into order those facts concerning the whole matter that I have been able to gather. Moreover, I have compiled this body of facts into a handbook to consult as a rule and guide for my administration. (3) In other books too, which I have written as a result of practical experience, provision was made for successors. Perhaps this notebook will also be useful to my successor in this post. But since it was written at the

start of my administration, it will be of most benefit for my own training and guidance.

3. So that I do not seem to have overlooked anything necessary for an understanding of the whole topic, I will first list the aqueducts which flow into Rome; then by whom and during whose consulship, how many years after the founding of the city, each was introduced; then from what place and what milestone it draws its water, how much is carried in an underground conduit, how much on substructure, and how much on arcade. (2) After the elevation of each, I will give an account of units of measurement, the distributions which have been made from them, how much each of the aqueducts supplies outside and inside the city, and in what measures to each region in relation to the supply available; how many *castella* there are, public and private,[1] and how much from them is supplied to public works and to *munera*—for that is the name given to more ornamental taps[2]—how much to basins, how much in Caesar's name, how much for the use of private citizens by grant of the emperor; what is the law for construction and maintenance of aqueducts and what penalties called for by statute, decree of the senate, and orders of the emperor sanction it.

4. From the foundation of the city for 441 years, the needs of Romans were satisfied by the water that they used to draw either from the Tiber, from wells, or from springs. (2) The memory of the springs is still considered holy and revered; indeed they are believed to restore sick bodies to health, such as the spring of the Camenae, and . . . that of Juturna.[3] (3) Now, however, there flow into the city the Aqua Appia, the Anio Vetus, the Marcia, the Tepula, the Julia, the Virgo, the Alsietina (which is also called the Augusta), the Claudia, and the Anio Novus.

---

1. The translation follows Rodgers' text, reading *castella publica privataque*. Private tanks were those distributing water to *privati*.

2. Cf. Bennett's comment on *munera* (334 n. 1): "the conventional interpretation of a very uncertain word." *Munera* are frequently private undertakings for public benefit, a term often applied to public gladiatorial shows offered by the emperor or other officials (so Pace, 237 n. 1). The idea of shows cannot apply here: Frontinus' listing of *munera* in his distribution statistics (78–86) indicates that they are to be thought of as fixed installations. The MS. tradition gives *munera—ita enim cultiores adpellantur,* prompting editors to supply a noun to be modified by the adjective. Kunderewicz gives *salientes* ("fountains" or "jets"). Jordan, 2:66, proposed *lacus* ("basins"), a reading adopted by Lanciani, 581; and Grimal, 65 n. 3.

3. The MS. tradition lists three springs: those of the Camenae and Juturna are well known, but the third, the *fons Apollinaris,* is otherwise unattested, and Frontinus' adjective is not parallel with the other two genitives. Jordan, 2:48, deleted the word as a gloss; Rodgers obelizes the text at this point.

5. In the consulship of Marcus Valerius Maximus and Publius Decius Mus [312 B.C.],[4] in the thirtieth year after the start of the Samnite War, the Aqua Appia was brought into the city by Appius Claudius Crassus the censor, who afterwards had the name Caecus. Appius also undertook construction of the Via Appia from the Porta Capena to the city of Capua. (2) He had as his colleague in the censorship Gaius Plautius, to whom the last name Venox was given, because he had searched for the veins [*venae*][5] from which the aqueduct was drawn. (3) But since Plautius resigned from the censorship within eighteen months, after he had been led to believe that his colleague Claudius would do the same, the glory of giving a name to the aqueduct went only to Appius; he is recorded to have drawn out his censorship by many maneuvers until he could complete both his highway and this aqueduct. (4) The Aqua Appia has its beginning on property once belonging to Lucullus along the Via Praenestina, between the seventh and eighth milestone, at 780 paces on a side road to the left. (5) Its conduit has a length of 11,190 paces from its source to the Salinae, which is an area near the Porta Trigemina; of this length there is an underground conduit for 11,130 paces and substructure and arcade next to the Porta Capena for 60 paces. . . . (6) Near Spes Vetus at the common boundary of the property of Torquatus and Taurus[6] there is joined to the Appia a supplemental branch, that of the Augusta added by Augustus; the watermen have given the name "Twins" to the place. (7) This branch takes its source at the sixth milestone on the Via Praenestina, on a side road 980 paces to the left, very close to the Via Collatina. (8) Its conduit to the "Twins" runs in an underground channel for 6,380 paces. (9) The Appia begins its distribution at the bottom of the Clivus Publicii at the Porta Trigemina, a place that is called the Salinae.

6. Forty years after the Appia was introduced, in the 481st year from the foundation of the city [272 B.C.], Manius Curius Dentatus, who carried out the censorship with Lucius Papirius Cursor, contracted—with spoils taken from Pyrrhus—for the introduction of the aqueduct that is now called the Anio Vetus, during the second consulship of Spurius Carvilius and Lucius Papirius. (2) Two years later the completion of this aqueduct was discussed in the senate. . . . praetor.[7] (3) Then by senatorial decree there were elected as a two-man commission

---

4. Modern dates have been added for all consular dates given in the text.

5. The pun cannot be exactly reproduced in translation, but I follow the rendering given by Ashby, 49.

6. The MS. tradition is corrupt here. Kunderewicz follows the reading *Tauriani* proposed by J. Carcopino, *La basilique pythagoracienne de la Porte Majeure* (Paris, 1943), 67–74. Cf. Ashby, 49 n. 9.

7. There is a gap in the text, perhaps indicating the procedure followed in appointment of the magistrates, with the name of the praetor who discussed the matter.

for completing the aqueduct Curius, who had issued the contracts for it, and Fulvius Flaccus. (4) Curius died within five days after he had been appointed one of the commissioners; the glory for completing the line went to Flaccus. (5) The Anio Vetus takes its source above Tibur at the twentieth milestone outside the Porta . . . ,[8] where it furnishes part of its capacity for the use of the Tiburtines. (6) Because of the requirements of leveling, its conduit has a length of 43,000 paces; of this there is an underground conduit of 42,779 paces and substructure above ground of 221 paces.

7. One hundred twenty-seven years later, in the 608th year after the foundation of the city, in the consulship of Servius Sulpicius Galba and Lucius Aurelius Cotta [144 B.C.], because the conduits of the Appia and Anio, already leaky from old age, were being diverted illegally by private parties, the senate gave Marcius, who then as praetor was administering justice between citizens, the task of rebuilding and reclaiming the lines. (2) In addition, because the growth of the city seemed to demand a more abundant supply of water, this same Marcius was also charged by the senate to take steps to introduce whatever additional water he could into the city. (3) Marcius rebuilt the earlier conduits and brought into the city a third aqueduct in its own conduit, which has the name Marcia from its builder.[9] (4) We read in Fenestella that the sum of 180,000,000 sesterces was allotted to Marcius for these projects, and since his term of office was not enough for completing the work, his praetorship was extended for a second year. (5) At that time the *decemviri sacris faciundis*, while consulting the Sibylline Books for other reasons, are said to have found that it was not right that the Aqua Marcia be brought to the Capitol, or rather the Anio Vetus—for the tradition is more consistent about the second aqueduct.[10] There was discussion about this question in the senate in the consulship of Appius Claudius and Quintus Caecilius [143 B.C.], when Marcus Lepidus spoke

8. The MS. reading is confused, and the location of the source of the Anio Vetus "at the twentieth milestone" does not agree with the archaeological evidence; see Ashby, 58–59. Blackman, "Length," 18–19, suggests that at least one line of text has been lost from the MSS. dealing with the real distance of the source east of Tibur and the location of the Tiburtine branch line. Rodgers in his forthcoming commentary suggests that Frontinus at 6.6 cites only the distance from Rome to Tibur, where the special branch line of the Anio Vetus was drawn to supply the Tiburtines. There is no convincing identification of the *porta* cited in 6.5; Lanciani, 460, arguing that the gate was one at Tibur, supports Cassio's emendation *Barana*. For more detailed discussion see Evans, "*In Tiburtium usum*," 450–51.

9. Cf. Rodgers, "Textual Temptations," 131.

10. This passage has been much discussed. For a summary of the problems and the text that this translation follows see R.H. Rodgers, "What the Sibyl Said: Frontinus *Aq*. 7.5," *CQ* 31 (1982): 174–77.

in behalf of the *decemviri,* and the same matter was brought up again three years afterwards by Lucius Lentulus, in the consulship of Gaius Laelius and Quintus Servilius [140 B.C.], but at each time the influence of Marcius Rex carried the day. In this way then the aqueduct was introduced to the Capitol.

(6) The Aqua Marcia begins at the thirty-sixth milestone on the Via Valeria, on a side road 3,000 paces to the right (for those coming from the city), but at the thirty-eighth milestone on the Via Sublacensis, which was first paved when Nero was emperor, to the left, less than 200 paces off the road. (7) . . It stands like a lake, with a very green color.[11] (8) Its conduit has a length from its intake to the city of 61,710½ paces; of this, 54,247½ paces run in an underground channel and 7,463 on structures above ground. Of this last portion, in many places farther from the city, 463 paces run across valleys on arcades, and closer to the city, from the seventh milestone, 528 paces run on substructure and the remaining 6,472 paces on arcade.

8. In the 627th year after the founding of Rome, in the consulship of Marcus Plautius Hypsaeus and Marcus Fulvius Flaccus [126 B.C.], the censors Gnaeus Servilius Caepio and Lucius Cassius Longinus, who was called Ravilla, took steps to bring to Rome and the Capitol the aqueduct that is called the Tepula from the property of Lucullus, which some believe to be part of Tusculan territory. (2) The Tepula begins at the tenth milestone on the Via Latina, 2,000 paces on a side road to the right (for those coming from Rome). From here it used to be brought in its own channel into the city.

9. Subsequently, as aedile after his first consulship, when the emperor Caesar Augustus was consul for the second time with Lucius Volcatius, in the 719th year after the founding of Rome [33 B.C.], Marcus Agrippa tapped the supply of another source at the twelfth milestone on the Via Latina, on a side road 2,000 paces to the right (for those coming from Rome) and cut off the conduit of the Tepula. (2) The name Aqua Julia was given to the new aqueduct by its builder, although its distribution was arranged in such a way that the name Tepula remained. (3) The conduit of the Julia runs a length of 15,426½ paces, 8,426 paces underground and 7,000 paces above ground.[12] Of this, in places closest to the city from the seventh milestone, 528 paces are on substructure and the remaining 6,472 on arcade.

(4) Beyond the intake of the Julia flows a stream that is called the Crabra. (5) Agrippa did not tap this, either because he had rejected its water or because

---

11. The text is corrupt here, but Frontinus appears to be describing conditions at the Marcia's source.

12. The notice of 8,426 *passus* underground appears in no MSS. but is added by Rodgers as a parallel to Frontinus' notices elsewhere.

he believed it should be left for the property owners of Tusculum, for it is this water that all the villas of that district receive in turn, distributed according to set days and amounts. (6) But our watermen, who do not use the same sort of restraint, have constantly claimed a part of that supply for a supplement to the Julia. They have done this, not to augment the delivery of the aqueduct, but because they used to drain the line by illicit distributions for their own profit. (7) The Crabra has therefore been shut off, and by order of the emperor has been wholly restored to the Tusculans, who now, perhaps not without wonder, make use of it, although they do not know to what cause they owe its unaccustomed abundance.[13] (8) Moreover, now that the illegal branches have been checked, through which it used to be robbed, the Julia has kept its normal capacity even in times of exceptional drought. (9) In the same year Agrippa restored the conduits of the Appia, Anio Vetus, and Marcia, which had almost worn out, and with extraordinary care provided the city with a great many taps.

10. After Agrippa had already been consul for the third time, in the consulship of Gaius Sentius and Quintus Lucretius [19 B.C.], thirteen years after he had built the Julia, he brought into Rome the Aqua Virgo, which he tapped on the property of Lucullus. (2) The day on which it first "declared its presence" to the city is recorded to have been June 9. (3) It was named the Virgo because a young maiden showed certain springs to soldiers hunting for water; after they had followed these springs and dug, they "summoned up" a huge supply of water.[14] (4) A shrine next to the source has a painting depicting this origin of the aqueduct. (5) The Virgo takes its source at the eighth milestone of the Via Collatina in a marshy area enclosed by a concrete barrier to contain the gushing springs. (6) Its capacity is also increased by a good number of other additional sources. (7) It comes to the city over a course of 14,105 paces, of which 12,865 paces run underground and 1,240 above ground. Of the course above ground, 540 paces are on substructure in many areas of cross-streams and 700 on arcade. (8) The underground supplementary conduits run a course of 1,405 paces.

11. What reason moved Augustus, an emperor with the greatest foresight, to introduce the Aqua Alsietina, which is also named the Augusta, I do not know. Its water enjoys no popularity; indeed it is hardly wholesome and is distributed nowhere for human consumption. Perhaps when Augustus began the task of building his naumachia, he brought this aqueduct expressly for it,

---

13. Cf. Rodgers, "Textual Temptations," 132–33.

14. Rodgers in his commentary on 10.2–3 suggests that Frontinus' special use of the verbs *respondere* and *vocare* may reflect colloquialisms found in Agrippa's own account of his aedileship.

to avoid drawing at all on the more beneficial water of other lines, and granted what began to be surplus for the naumachia to nearby properties and to the use of private parties for irrigation. (2) On the other hand, distribution of it is customarily made in the Transtiber region, whenever bridges are being repaired and water distribution from the near side of the river is interrupted, as an emergency supplement to public taps. (3) The Aqua Alsietina takes its source from the Lacus Alsietinus, at the fifteenth milestone of the Via Claudia on a side road 6,500 paces to the right. (4) Its conduit runs a length of 22,172 paces, with 358 paces on arcade.

12. To supplement the Aqua Marcia whenever periods of drought required an additional supply, Augustus also added to it by an underground conduit another source of water of the same quality, which is named the Augusta from its builder. (2) This branch begins beyond the source of the Marcia. Its conduit, before it reaches that of the Marcia, runs 800 paces.

13. After these lines, since seven aqueducts appeared inadequate for meeting public needs and the personal demands of private users, Gaius Caesar, who succeeded Tiberius, in the second year of his principate, during the consulship of Marcus Aquila Julianus and Publius Nonius Asprenas, the 789th year after the founding of Rome [A.D. 38], began construction of two aqueducts. (2) Claudius brought this project to a most splendid completion and dedicated it in the consulship of Faustus Sulla and Salvius Otho,[15] the 803d year after the founding of Rome [A.D. 52], on August 1. (3) The name Aqua Claudia was given to the one aqueduct that was brought from the Caerulan and Curtian springs. (4) It is next to the Marcia in quality. (5) The other began to be called the Anio Novus, since two aqueducts supplied by the Anio River began to flow into the city, so that they might be distinguished more easily by their titles. The Anio Novus takes precedence in height over all other aqueducts. The name Vetus was added to the earlier Anio line.

14. The Aqua Claudia takes its source at the thirty-eighth milestone of the Via Sublacensis, within 300 paces on a side road to the left, from two very abundant and beautiful springs, the Caerulan, which takes its name from its appearance, and the Curtian. (2) It also receives that spring which is called the Albudine, one of such high quality that whenever the Marcia also needs supplementing, the Albudine is so suitable that it changes nothing of that aqueduct's quality by being added to it. (3) The source of the Aqua Augusta has been diverted into the Claudia, since it was evident that the Marcia itself was of sufficient volume, but it remains nonetheless an auxiliary supply for the Marcia,

15. Cf. Rodgers, "Textual Temptations," 133.

in such a way that it supplies the Aqua Claudia only if the conduit of the Marcia does not receive it. (4) The Claudia has a length of 46,406 paces, of which 36,230 paces run underground and 10,176 are above ground. Of that above ground, there are 3,076 paces in different places along its upper course, and near the city from the seventh milestone, 609 paces on substructure and 6,491 on arcade.

15. The Anio Novus has its intake at the forty-second milestone on the Via Sublacensis, in Simbruine territory, from the Anio River, which flows muddy and turbid even without the bad effect of rains, since it has cultivated and rich fields around it and, as a result, quite loose banks. (2) For this reason a settling tank was installed away from the intake of the conduit, where the water might settle and be filtered between the river and aqueduct channel. (3) But even so, it comes to the city turbid whenever there are heavy rains. (4) There is joined to it the Herculean brook, which arises at the thirty-eighth milestone on the same road from the region of the Claudia's sources, on the other side of the river and highway. (5) It is very pure at its source, but after being mixed with the Anio it loses the charm of its own limpidity. (6) The Anio Novus has a course of 58,700 paces, of which 49,300 paces run underground and 9,400 above ground. Of that above ground, 2,300 paces are on substructures or arcade in various places along its upper course. Closer to the city, from the seventh milestone, there are 609 paces on substructure and 6,491 paces on arches. These are the highest arches, raised in certain places to a level of 109 feet.

16. With so many indispensable structures carrying so many aqueducts you may compare the idle pyramids or the other useless, although famous, works of the Greeks!

17. It has seemed to me not unfitting to include as well a description of the lengths of the courses of each aqueduct, according to the classifications of construction. (2) Because the greatest part of the duties of this position lies in maintenance of the lines, the man in charge must know what things demand greater outlays. (3) My sense of responsibility has not been satisfied with personal examination of particular items. I have also taken care to prepare maps of the lines, from which it is clear where there are valleys and how great they are, where rivers are crossed, and where channels attached to the sides of mountains demand greater and constant attention for their . . . repair.[16] (4) As a result, we gain the advantage of being able to deal with a matter immediately, as if under our own eyes, and to make a decision as though we were on the spot.

---

16. The text is unclear, but Frontinus' phrase seems to have been something like "maintenance and repair." Rodgers compares Frontinus' more detailed comments in chapter 121.

18. All the aqueducts arrive in the city at different elevations. (2) As a result, certain ones serve higher places, and others cannot be raised to more lofty areas; indeed even the hills have grown up little by little from rubble on account of the great number of fires. (3) The height of five aqueducts permits them to be raised into every part of the city, but of these, some are forced by greater pressure, others by less. (4) The highest of all is the Anio Novus, the next highest, the Claudia; the Julia holds third place, the fourth the Tepula, and after this the Marcia, which even equals the height of the Claudia at its source. But the old aqueduct builders constructed their lines at a lower elevation, either because the fine points of the leveling art had not yet been ascertained, or because they deliberately made it their practice to bury aqueducts underground to prevent them from being cut easily by enemies, since a good many wars were still being fought against the Italians. (5) Now, however, in certain places, whenever a conduit has been worn out by age, the lines are rebuilt to be carried across valleys on substructures or arches to shorten their length and avoid underground detours around the low ground. (6) The Anio Vetus holds sixth place in height; it would in like manner supply even the higher districts of the city if it were raised on substructures and arcades where the nature of valleys and low ground demands. (7) Next follows the Virgo, then the Appia; since these lines were brought from territory near the city, they could not be raised to so great an altitude. (8) The lowest of all is the Alsietina, which serves the Transtiber region and, most of all, low-lying places.

19. Six of these lines are received in covered settling tanks on the Via Latina, this side of the seventh milestone, where their channels take fresh breath from their journey, so to speak, and they deposit their sediment. (2) Their volume is also determined by measuring gauges installed there. (3) Moreover, these three lines, the Julia, Marcia, and Tepula, come out of the ground together.[17] (The Tepula, as we indicated earlier, which had been tapped and added to the conduit of the Julia, has now received a portion of the Julia's water from the settling tank of that aqueduct and enters Rome in its own channel and under its own name.) These three lines are carried from their settling tanks onto the same arches. (4) The highest is the Julia, below it, the Tepula, then the Marcia. (5) Continuing to run together to the level of the Viminal Hill, they reach the Porta Viminalis. (6) There they again come out of the ground. (7) Part of the Julia, however, drawn off near Spes Vetus, is distributed through *castella* on the Caelian Hill. (8) In addition, the Marcia casts off part of its volume behind the

---

17. The translation follows Rodgers' emendation *emergunt* for the unintelligible *earum* of the MS. tradition: *emergunt* is consistent with the archaeological evidence at Capannelle and makes Frontinus' statement *rursus emergunt* (19.6) much more logical.

estate of Pallas into a channel that is called the *rivus Herculaneus*. (9) Running across the Caelian but supplying no needs of that hill because of its low level, this branch terminates above the Porta Capena.

20. The Anio Novus and the Claudia are carried from their settling tanks on a single arcade higher than that of the Marcia, Tepula, and Julia, with the Anio Novus on top. (2) Their arches end behind the estate of Pallas, and from here their water is distributed in pipes to supply the city's needs. (3) The Claudia, however, delivers part of its volume beforehand at Spes Vetus to the so-called Neronian Arches. (4) This arcade runs over the Caelian Hill and terminates next to the Temple of the Divine Claudius. (5) The Neronian Arches distribute the volume they have received either around the Caelian itself or to the Palatine, Aventine, and Transtiber.

21. The Anio Vetus itself also has a settling tank on this side of the fourth milestone, on a path that crosses between the arches from the Via Latina to the Via Labicana.[18] (2) From here, inside the second milestone, it gives part of its volume to a channel called the Octavianus, which runs to the region of the Via Nova near the estate of Asinius, from where it is distributed throughout that region. (3) The main aqueduct, however, coming behind Spes Vetus inside the Porta Esquilina, is distributed into deep channels throughout the city.

22. Neither the Virgo, the Appia, nor the Alsietina has catch basins, that is, settling tanks. (2) The arcade of the Virgo begins at the foot of the Lucullan estate and ends in the Campus Martius after passing in front of the Saepta. (3) The Appia's channel, having run under the Caelian and Aventine, comes out of the ground, as we have said, below the Clivus Publicii. (4) The conduit of the Alsietina terminates behind the naumachia, for which it seems to have been built.

23. Since I have reviewed the builders and dates of each aqueduct, in addition to their sources, the lengths of their conduits, and the order of their elevations, it seems to me not out of place to add individual details as well and to show how great is the supply that satisfies not only public and private needs and support but also pleasures; through how many *castella* and to what regions water is distributed, how much outside the city, how much within it, and how much of this volume is furnished to basins, how much to ornamental fountains, how much to public works, how much in Caesar's name, how much for private needs. (2) I think it worthwhile, however, before citing the names of *quinariae*, *centenariae*, and the other units through which measurement has been estab-

---

18. The MSS. are corrupt here. I follow the suggestion made by Rodgers: *in tramite qui a via Latina in Lavicanam traicit et ipse piscinam habet*. For other suggestions cf. Grimal, 79 n. 57; Pace, 254 n. 2; and Lanciani, 271, followed by Ashby, 55 n. 6.

lished, to point out what their origin is, what their capacities are, and what each term means, and after setting forth the rule according to which their system is reckoned and fixed, to show how I have found things in disagreement with this system and what way of correcting them I have followed.

24. Units of water measurement have been established according to digits or inches. That of the digit is followed in Campania and most places of Italy; inches are still followed in Apulia.[19] (2) A digit, moreover, is agreed to be the sixteenth part of a foot, an inch the twelfth part. (3) But unlike the difference between the inch and the digit, there is a double rule for the digit itself.[20] (4) One type is called the square digit, another the round digit. (5) The square digit is larger than the round one by $^3/_{14}$, the round digit smaller than the square one by $^3/_{11}$, precisely because the corners are subtracted.

25. At a later period another unit of measure developed, which is called the *quinaria* or 5-pipe, taking its origin neither from the inch nor from either type of digit. It was introduced by Agrippa, as some people think, or, according to others, by lead manufacturers under the direction of the architect Vitruvius, and it has come into general use in Rome, replacing the other units of measure. (2) Those who say Agrippa was its inventor derive its name from the fact that there were combined into one pipe five old units, small and like points, through which water used to be distributed, since it was in small quantity. Those who attribute the *quinaria* to Vitruvius and the lead workers say that is because a flat sheet of lead having a width of five digits produces a pipe of this unit of measurement when formed into a cylinder. (3) This, however, is uncertain, because when a cylinder is formed, just as the sheet is contracted on its inside surface, it is enlarged on that surface which is outside. (4) The most plausible explanation of the name *quinaria* is that it is derived from its diameter of five quarter-digits, a system that is maintained in units of measure that follow, up to the *vicenaria,* or 20-pipe, the diameter in each unit increasing by the addition of individual quarter-digits, as in the *senaria,* or 6-pipe, which has a diameter of six-fourths, and the *septenaria,* or 7-pipe, which has seven, and so on by similar increases, up to the *vicenaria.*

26. Moreover, every unit of measurement is calculated either by its diameter, its perimeter, or by the dimensions of its cross section; from these calculations its capacity is made evident. (2) To distinguish more easily the difference between units of inches, square digits, and round digits, as well as that of the *quinaria* itself, we must use the *quinaria* as our standard, since it is the unit that is most fixed and most widely in use. (3) An inch unit of measurement therefore has

19. Cf. Rodgers, "Textual Temptations," 133.
20. Cf. Rodgers, "Textual Temptations," 133.

a diameter of $1\frac{1}{3}$ digits; it receives more than $1\frac{1}{8}$ *quinariae*, that is, $1\frac{1}{2}$ twelfths more, plus $\frac{3}{288}$ plus two-thirds of $\frac{1}{288}$. (4) A square digit reduced to a circle has a diameter of one digit plus $1\frac{1}{2}$ twelfths plus $\frac{1}{72}$; it has a capacity of $\frac{5}{6}$ of a *quinaria*. (5) A round digit has a diameter of one digit, with a capacity of $\frac{7}{12}$ plus one twenty-fourth plus $\frac{1}{72}$ of a *quinaria*.

27. On the other hand, those units of measurement that are based on the *quinaria* take their increase in two ways. (2) One method employed is when the *quinariae* themselves are doubled or multiplied,[21] that is, when more than one *quinaria* is contained within the same opening, in which the size of the opening increases according to the number of *quinariae* added. (3) This is the method generally used whenever more than one *quinaria* that have been procured are delivered through one pipe to a *castellum* from which individual pipes receive their own allotment, so that the flow of the conduit not be weakened by too many taps.

28. The second method is used whenever the pipe takes its increase, not according to a fixed number of *quinariae*, but instead by the measure of its own diameter, according to which it both takes its name and enlarges its capacity. For example, a *quinaria* pipe, when a quarter-digit has been added to its diameter, becomes a *senaria* or 6-pipe. (2) It does not increase its capacity by a complete *quinaria*, since the pipe when enlarged takes one *quinaria* plus $\frac{5}{12}$ plus $\frac{1}{48}$. (3) By this same method, too, as quarter-digits are added to the diameter, the result, as indicated above, is a *septenaria* or 7-pipe, and *octonaria* or 8-pipe, and so on, up to a *vicenaria*, or 20-pipe.

29. From this point another method of calculation is observed, one based on the number of square digits that are contained in the surface or opening of each unit of measurement, from which the pipes also take their name. (2) A pipe that has an opening of 25 square digits is called a *vicenum quinum*, or 25-pipe. There follow in like manner the *tricenaria* or 30-pipe, and so on, in increments of five square digits, up to the *centenum vicenum* or 120-pipe.

30. In the case of *vicenaria* or 20-pipe, which is on the border between the two methods of calculation, both methods practically coincide. (2) This pipe, after all, has a diameter of 20 quarter-digits according to the reckoning that must be followed in smaller units, since it has a diameter of 20 square digits, minus a small amount.

31. This method of calculating pipe capacity by *quinariae* is followed through all units of measurement up to the 120-pipe, as we have shown, and is consistent in its application to every class. (2) It is also in agreement with those standards

21. Cf. Rodgers, "Textual Temptations," 133–34.

that have been entered and verified in the records of our most mighty and devoted emperor. (3) Whether the system then or the emperor's authority is to be followed, the units of measurement in the records take precedence. (4) But the watermen, although they conform to the standard computation in the case of most pipes, have modified it in four units, the 12-pipe, 20-pipe, 100-pipe, and 120-pipe.

32. In the case of the 12-pipe, to be sure, neither the error is great nor is the practice common. (2) The watermen have added to its diameter $\frac{1}{24}$ plus $\frac{1}{48}$ of a digit and to its capacity . . . of a *quinaria*. . . .[22] (3) In the other three units, however, there is a greater discrepancy. (4) They make the 20-pipe smaller in diameter by a half-digit and in capacity by three *quinariae* plus $\frac{1}{24}$. (5) They commonly make distribution through pipes of this measurement. (6) In the case of the 100- and 120-pipes, however, through which the watermen constantly receive water for later distribution, they do not diminish but increase the units.[23] (7) For they add $\frac{2}{3}$ of a digit plus $\frac{1}{24}$ to the diameter of a 100-pipe and 10 *quinariae* plus $\frac{2}{3}$ plus $\frac{1}{24}$ to its capacity. (8) In the case of the 120-pipe, they add three digits plus $\frac{7}{12}$ plus $\frac{1}{24}$ plus $\frac{1}{48}$ to the diameter, and to its capacity they add 66 *quinariae* plus $\frac{1}{6}$.

33. In this way, while they decrease the capacity of the 20-pipe, through which they frequently distribute, or add to that of the 100- or 120-pipes, by which they always receive water for subsequent distribution, there are illegally tapped 27 *quinariae,* in the case of the 100-pipe, and 86 *quinariae,* in the case of the 120-pipe. (2) Although this practice can be proven with calculations, it is also obvious from the very fact. (3) For with respect to the 20-pipe, which Caesar's records list as supplying 16 *quinariae,* the watermen distribute no more than 13, and from a 120-pipe, it is clear that they supply only a more limited amount, since Caesar, according to his own record books, curtails distribution, as if the capacity were exhausted, when he has supplied 81$\frac{1}{2}$ *quinariae* from a 100-pipe and likewise 98 *quinariae* from a 120-pipe.

34. To sum up: there are 25 units of measurement for pipes. (2) All of them are in agreement with our calculations and the official records except these four that the watermen have altered. (3) All standards connected with measurement, however, ought to be fixed and internally consistent, for in this way the system will have general application. (4) For example, as a pint has its own ratio to cup measures, but peck measures correspond to both pints and cups, so *quinariae,* when multiplied in larger units, ought to follow a rule of their progression. (5)

---

22. The text is confused, and Frontinus seems to indicate a small quantity of a *quinaria* added to the volume, although he may have been more precise.

23. Cf. Rodgers, "Textual Temptations," 134.

Otherwise when a loss is discovered in a unit of measurement that the watermen supply and gain in that which they generally receive, it is evident that there is no error but fraud.

35. We remember that all water, whenever it comes from a higher elevation and falls into a *castellum* within a short distance, not only corresponds to its original unit of measurement but also exceeds it; on the other hand, whenever it is brought from a lower elevation, that is, under less pressure, for a longer distance, it also decreases its delivery because of the slowness of its passage. For this reason then, according to this reckoning, its delivery figures must be supplemented or decreased.

36. But the position of the delivery fitting also has an important effect.[24] (2) In a straight and level position the spigot maintains a constant measure of water. If it is set downward and against the flow of the water, that is, on an incline toward the intake, it draws more. On an incline upward and turned toward the flow of the current, it receives less water and at a slower rate. (3) A delivery fitting is a bronze nozzle that is installed in a conduit or *castellum*. It is to this that pipes are attached. (4) A fitting ought to have no less than twelve digits of length with a cross section, that is, a capacity, as great as has been granted. (5) The choice of material for it seems to have been carefully thought out, since bronze, being hard and difficult to bend, cannot be accidentally expanded or contracted.

37. I have listed below the dimensions of the units of measurement, 25 in number, although only 15 are in general use, set forth according to the system about which we spoke earlier, with corrections made in the case of the four measures altered by the watermen. (2) All pipes that will be in use ought to be brought into conformity with this system, or, if these particular pipes remain, their capacities ought to be calculated according to the number of *quinariae* they receive. (3) Those units not in use are so noted.[25]

39. The *quinaria* pipe: diameter, 1 digit plus 1/4; circumference, 3 digits plus 11/12 plus 3/288 [3.9272 digits];[26] capacity, 1 *quinaria*.

40. The *senaria* or 6-pipe: diameter, 1 digit plus 1/2; circumference, 4

24. Frontinus uses the term *calix,* which normally means "drinking cup." For these delivery fittings see Hodge, 294–300. On the problem of the *calix* as described in this chapter of Frontinus see Bruun, 41–44.

25. Chapter 38 in the MSS. repeats the same points about pipe measurement found in chapter 26; editors have therefore deleted it as a later gloss. For one explanation of how chapter 38 intruded into the MS. tradition see Rodgers, "Textual Tradition," 102–3.

26. Decimal equivalents of multiple fractions have been added to the translation for clarity.

digits plus 2/3 plus 1/24 plus 2/288 [4.7153 digits]; capacity, 1 *quinaria* plus 5/12 plus 7/288 [1.4410 *quinariae*].

41. The *septenaria* or 7-pipe: diameter, 1 digit plus 3/4; circumference, 5 1/2 digits; capacity, 1 *quinaria* plus 11/12 plus 1/24 [1.9583 *quinariae*]; not in use.
42. The *octonaria* or 8-pipe: diameter, 2 digits; circumference, 6 digits plus 1/4 plus 10/288 [6.2847 digits]; capacity, 2 *quinariae* plus 1/2 plus 1/24 plus 5/288 [2.5590 *quinariae*].
43. The *denaria* or 10-pipe: diameter, 2 digits plus 1/2; circumference, 7 digits plus 5/6 plus 7/288 [7.8576 digits]; capacity, 4 *quinariae*.
44. The *duodenaria* or 12-pipe: diameter, 3 digits; circumference, 9 digits plus 5/12 plus 3/288 [9.4271 digits]; capacity, 5 *quinariae* plus 3/4 plus 3/288 [5.7604 *quinariae*]; not in use. (2) After being altered by the watermen it had a diameter of 3 digits plus 1/24 plus 6/288 [3.0625 digits] and a capacity of 6 *quinariae*.
45. The 15-pipe: diameter, 3 digits plus 3/4; circumference, 11 digits plus 3/4 plus 10/288 [11.7847 digits]; capacity, 9 *quinariae*.
46. The *vicenaria* or 20-pipe: diameter, 5 digits; circumference, 15 digits plus 2/3 plus 1/24 plus 2/288 [15.7153 digits]; capacity, 16 *quinariae*. (2) After being altered by the watermen it had a diameter of 4 1/2 digits and a capacity of 12 *quinariae* plus 11/12 plus 1/24 [12.9583 *quinariae*].
47. The 25-pipe: diameter, 5 digits plus 7/12 plus 1/24 plus 5/288 [5.6424 digits]; circumference, 17 digits plus 2/3 plus 1/24 plus 6/288 [17.7292 digits]; capacity, 20 *quinariae* plus 1/3 plus 9/288 [20.3646 *quinariae*]; not in use.
48. The *tricenaria* or 30-pipe: diameter, 6 digits plus 1/6 plus 4/288 [6.1806 digits]; circumference, 19 digits plus 5/12 [19.4166 digits]; capacity, 24 *quinariae* plus 5/12 plus 5/288 [24.4340 *quinariae*].
49. The 35-pipe: diameter, 6 digits plus 2/3 plus 2/288 [6.6736 digits]; circumference, 20 digits plus 11/12 plus 1/24 plus 5/288 [20.9757 digits]; capacity, 28 *quinariae* plus 1/2 plus 3/288 [28.5104 *quinariae*]; not in use.
50. The *quadragenaria* or 40-pipe: diameter, 7 digits plus 1/8 plus 3/288 [7.1354 digits]; circumference, 22 digits plus 5/12 [22.4167 digits]; capacity, 36 *quinariae* plus 7/12 [36.5833 *quinariae*].
51. The 45-pipe: diameter, 7 digits plus 1/2 plus 1/24 plus 8/288 [7.5694 digits]; circumference, 23 digits plus 3/4 plus 10/288 [23.7847 digits]; capacity, 36 *quinariae* plus 7/12 plus 1/24 plus 8/288 [36.6528 *quinariae*]; not in use.
52. The *quinquagenaria* or 50-pipe: diameter, 7 digits plus 11/12 plus 1/24 plus 5/288 [7.9760 digits]; circumference, 25 digits plus 1/24 plus 9/288 [25.0729 digits]; capacity, 40 *quinariae* plus 2/3 plus 1/24 plus 5/288 [40.7257 *quinariae*].

53. The 55-pipe: diameter, 8 digits plus 1/3 plus 10/288 [8.3681 digits]; circumference, 26 digits plus 1/4 plus 1/24 [26.2917 digits]; capacity, 44 *quinariae* plus 3/4 plus 1/24 plus 2/288 [44.7986 *quinariae*]; not in use.
54. The *sexagenaria* or 60-pipe: diameter, 8 digits plus 2/3 plus 1/24 plus 9/288 [8.7396 digits]; circumference, 27 digits plus 5/12 plus 1/24 [27.4583 digits]; capacity, 48 *quinariae* plus 5/6 plus 11/288 [48.8715 *quinariae*].
55. The 65-pipe: diameter, 9 digits plus 1/12 plus 3/288 [9.0938 digits]; circumference, 28 digits plus 7/12 [28.5833 digits]; capacity, 52 *quinariae* plus 11/12 plus 8/288 [52.9444 *quinariae*]; not in use.
56. The *septuagenaria* or 70-pipe: diameter, 9 digits plus 5/12 plus 6/288 [9.4375 digits]; circumference, 29 digits plus 2/3 [29.6667 digits]; capacity, 57 *quinariae* plus 5/288 [57.0174 *quinariae*].
57. The 75-pipe: diameter, 9 digits plus 3/4 plus 6/288 [9.7709 digits]; circumference, 30 digits plus 2/3 plus 1/24 [30.7083 digits]; capacity, 61 *quinariae* plus 1/12 plus 2/288 [61.0903 *quinariae*]; not in use.
58. The *octogenaria* or 80-pipe: diameter, 10 digits plus 1/12 plus 2/288 [10.0903 digits]; circumference, 31 digits plus 2/3 plus 1/24 [31.7083 digits]; capacity, 65 *quinariae* plus 1/6 [65.1667 *quinariae*].
59. The 85-pipe: diameter, 10 digits plus 1/3 plus 1/24 plus 7/288 [10.3993 digits]; circumference, 32 digits plus 2/3 plus 6/288 [32.6875 digits]; capacity, 69 *quinariae* plus 1/6 plus 1/24 plus 8/288 [69.2361 *quinariae*]; not in use.
60. The *nonagenaria* or 90-pipe: diameter: 10 digits plus 2/3 plus 10/288 [10.7014 digits]; circumference, 33 digits plus 7/12 plus 1/24 plus 3/288 [33.6354 digits]; capacity, 73 *quinariae* plus 1/4 plus 1/24 plus 5/288 [73.3090 *quinariae*].
61. The 95-pipe: diameter, 10 digits plus 11/12 plus 1/24 plus 11/288 [10.9965 digits]; circumference, 34 digits plus 1/2 plus 1/24 plus 5/288 [34.5590 digits]; capacity, 77 *quinariae* plus 1/3 plus 1/24 plus 2/288 [77.3819 *quinariae*]; not in use.
62. The *centenaria* or 100-pipe: diameter, 11 digits plus 1/4 plus 9/288 [11.2813 digits]; circumference, 35 digits plus 5/12 plus 1/24 [35.4583 digits]; capacity, 81 *quinariae* plus 5/12 plus 10/288 [81.4514 *quinariae*]. (2) After being altered by the watermen it had a diameter of 12 digits and capacity of 92 *quinariae* plus 1/12 plus 1/24 plus 10/288 [92.1597 *quinariae*].
63. The 120-pipe: diameter, 12 digits plus 1/3 plus 7/288 [12.3576 digits]; circumference, 38 digits plus 5/6 [38.8333 digits]; capacity, 97 *quinariae* plus 3/4 [97.75 *quinariae*]. (2) After being altered by the water-

men it had a diameter of 16 digits and a capacity of 163 *quinariae* plus $^{11}/_{12}$ [163.9167 *quinariae*], the equivalent of two 100-pipes.[27]

64. Having reviewed those things that needed to be said about units of measurement, I shall now set forth what capacity each aqueduct seemed to have, according to the imperial record books, until the time of my assumption of office, and how much it distributed, then the actual capacity that we ourselves discovered by careful investigation, with the foresight of Nerva, the best and most diligent of emperors, as our inspiration. (2) There were therefore a total of 12,755 *quinariae* according to the record books, and 14,018 *quinariae* in distribution, 1,263 more in distribution than the figure calculated as received. (3) Since I believed it the special task of my office to ascertain the reliability and actual capacity of the water supply, my wonderment at such a discrepancy impelled me to search out how more was being distributed than there was in the inheritance, so to speak. (4) First of all, I undertook to measure the intakes of the lines, but I found a capacity far greater than that listed in the record books, that is, a difference of about 10,000 *quinariae,* as I will show in individual discussions.

65. For the Appia, a capacity of 841 *quinariae* was listed in the records. (2) Measurement of this line could not be taken at its intake, since it consists of two channels. (3) But at the "Twins," a place below Spes Vetus, where the Appia is joined to the branch of the Augusta, I found the depth of water to be 5 feet and its width 1¾ feet, making 8¾ square feet in area, equal to 22 100-pipes and a 40-pipe, which results in 1,825 *quinariae,* 984 more than it has listed in the record books. (4) The Appia used to deliver 704 *quinariae,* 137 fewer than listed in the record books and, moreover, 1,121 fewer than that indicated by the measurements taken at the "Twins." (5) Some part of this loss is to be attributed to the poor condition of the conduit; since it runs at a deep level, it actually does not disclose its leaks. It is evident, however, that leaks do exist, because, in many sections of the city, water of good quality is observed. (6) But we have also caught certain illegal taps within the city. (7) Outside the city, however, on account of the gravitational flow of its level (since it is 50 feet below ground at its source), it receives no such damage.

66. For the Anio Vetus there was a capacity of 1,541 *quinariae* listed in the record books. (2) At its intake I found 4,398 *quinariae,* not including the amount taken off for the special conduit of the Tiburtines, 2,857 *quinariae* more than in the records. (3) Before it reached its settling tank, 262 *quinariae*

---

27. Several MSS. begin a new book at this point, but Frontinus always refers to his treatise in the singular. Cf. Grimal, xx–xxi; Kunderewicz, xi.

were distributed. (4) The capacity in the settling tank, which is calculated by the gauges installed there, is 2,362 *quinariae*. (5) There were lost therefore between the intake and the settling tank 1,774 *quinariae*. (6) Below its settling tank the Anio Vetus used to deliver 1,348 *quinariae*, 69 more than we stated were indicated as its initial capacity in the record books, and 1,014 fewer than we determined were received into the conduit below the settling tank. (7) The total of that lost between the intake and settling tank and below the settling tank is 2,788 *quinariae*, which I would suspect to have resulted from error in measurement had I not discovered where they were being diverted.

67. For the Marcia, there was a capacity of 2,162 *quinariae* listed in the record books. (2) Having measured at its intake, I found there 4,690 *quinariae*, 2,528 more than in the records. (3) There were distributed before it reached its settling tank 95 *quinariae*. In addition, 92 *quinariae* were also given as a supplement to the Tepula, as well as 164 to the Anio. (4) The total delivered before the settling tank was 351 *quinariae*. (5) The capacity, as calculated by the gauges installed in the settling tank, together with the amount which, carried around the settling tank, is received in the same channel on arches, results in 2,944 *quinariae*. (6) The total, either distributed before the settling tank or received on arches, is 3,295 *quinariae*, 1,133 more than that set down in the registers of the record books and 1,395 fewer than that indicated by the measurements at the intake. (7) The Marcia used to distribute below its settling tank 1,840 *quinariae*, 227 fewer than the capacity we said was indicated in the registers and 1,104 fewer than that received on arches from the settling tank. (8) The complete total lost, either between the intake and settling tank or below the settling tank, is 2,499 *quinariae*, a volume which, as in other lines, we discovered was illegally tapped in various places. (9) Indeed it is evident that there is no lack of water from the fact that at its intake, in addition to that volume which we set down as included in the capacity of the conduit, there is an overflow of more than 300 *quinariae*.[28]

68. For the Tepula, there was a capacity of 400 *quinariae* listed in the record books. (2) There are no springs belonging to this aqueduct: it used to arise from certain rivulets that have been tapped for the Julia. (3) Its intake is therefore to be recognized as the settling tank of the Julia. (4) From this line it first receives 190 *quinariae*, then immediately afterwards from the Marcia 92 *quinariae*, and in addition 163 from the Anio Novus near the estate of Epaphroditus. (5) Altogether there are 445 *quinariae* that appear in its distribution, 45 more than in the record books.

---

28. Cf. Rodgers, "Textual Temptations," 134.

69. For the Julia, there was a capacity of 649 *quinariae* listed in the record books. (2) Its capacity could not be determined at its source,[29] since it consists of more than one tributary, but it is received altogether into a settling tank at the sixth milestone from the city, where its capacity, according to clear measurements, is 1,206 *quinariae*, 557 more than in the record books. (3) Moreover, close to the city behind the estate of Pallas it receives 162 *quinariae* from the Claudia. (4) The total received by the Julia is 1,368 *quinariae*. (5) From this amount it supplies 190 *quinariae* to the Tepula and delivers 803 in its own name. (6) The total it distributes is 993 *quinariae*, 344 more than it has according to the record books, and 213 fewer than we determined it has in its settling tank, the very amount we discovered tapped by people without authorization of the emperor.

70. For the Virgo, there was a capacity of 652 *quinariae* listed in the record books. (2) It was not possible to measure its capacity at its intake, since it is drawn from several tributaries and enters its conduit too slowly. (3) Close to the city, however, at the seventh milestone, on property now belonging to Ceionius Commodus, where it has a swifter current, I took a measurement amounting to 2,504 *quinariae*, 1,852 more than in the record books. (4) Our confirmation is very much at hand, for the Virgo delivers all the volume we determined by measurement, that is, 2,504 *quinariae*.

71. The capacity of the Alsietina has neither been listed in the record books nor could it be determined with accuracy under present conditions, since it receives from the Lacus Alsietinus and then from the Lacus Sabatinus in the vicinity of Careiae as much water as the watermen have arranged. (2) The Alsietina delivers 392 *quinariae*.

72. The Claudia, being more plentiful than the others, has been most exposed to illegal taps. (2) It has a capacity in the record books of not more than 2,855 *quinariae*, although I found at its intake 4,607 *quinariae*, 1,752 more than in the records. (3) Our measurement, however, is secure to such a degree that in the settling tank at the seventh milestone from the city, where the measurements are not to be questioned, we find 3,312 *quinariae*, 457 more than in the record books, although it does make authorized distributions before it reaches this settling tank, and we have also discovered that a very large amount is illegally diverted. For that reason too there are found 1,295 fewer *quinariae* than there actually ought to be. (4) However, graft is also evident in its distribution, since the figures concerning it agree neither with the reliability of the record books nor with the measurements we made at the intake, nor even with

29. Cf. Rodgers, "Textual Temptations," 134.

those that were taken at the settling tank after so many illegal taps. (5) For only 1,750 *quinariae* are delivered, 1,105 fewer than the schedule of the record books indicates, 2,857 fewer than the measurements at the intake have shown, and even 1,562 fewer than the volume found in the settling tank. (6) For this reason, although it reaches the city separately in its own conduit, it used to be mixed there with the Anio Novus, with the result that from this combination of waters their individual capacity and distribution remained unknown. (7) But if some people think I am exaggerating in my measurements of its capacity, they should be reminded that the Curtian and Caerulan springs are so abundant in supplying to their conduit the 4,607 *quinariae* that I have indicated that there is an overflow of 1,600 excess *quinariae*. (8) I do not deny that this excess water does not properly belong to these springs; it is in fact taken from the Augusta branch devised as a supplement to the Marcia. We have added this branch to the springs of the Claudia whenever the Marcia does not need it, although there is so much water available that even the Claudia's conduit cannot receive all of it.

73. The Anio Novus used to be scheduled in the record books as having 3,263 *quinariae*. (2) Having measured it at its intake, I found 4,738 *quinariae*, 1,475 more than in the record books. (3) In what other way may I more obviously prove that I do not exaggerate this capacity than by the fact that the greater part of it is reflected in the distribution figures of the record books themselves? (4) According to them, 4,200 *quinariae* are delivered, although elsewhere in the same records a capacity of not more than 3,263 *quinariae* is found. (5) Moreover, I have discovered that there are illegally tapped not only 538 *quinariae*, the difference between our measurements and the scheduled delivery, but a much larger amount. (6) As a result, it is evident that the actual volume is even in excess of our measurements. This can be explained because the rather swift force of the water, taken from a bountiful and fast-flowing river, increases its volume by its very speed.

74. I have no doubt that some will be surprised that a far greater supply has been found by measurements we have taken than was indicated in the imperial record books.[30] (2) The reason for this is the error of those who originally made an estimate of the volume of each line with too little attention. (3) There is no reason to believe that they fell so far short of the true volume because of fear of summer or dry spells; I myself have discovered that the volume of each line indicated above, taken from measurements made in the month of July, remained constant afterwards throughout the whole summer.

---

30. Cf. Rodgers, "Textual Temptations," 134.

(4) Yet, whatever the earlier cause of the discrepancy, the following certainty has been revealed: 10,000 *quinariae* have been lost, while the emperors have limited their grants according to the volume listed in the record books.

75. A subsequent discrepancy is the result of one system of measurement being used at the intakes, another much smaller in the settling tanks, and another, the smallest of all, maintained in the delivery system. (2) The reason for this is the dishonesty of the watermen who we discovered were diverting water from public conduits for the use of private parties. (3) Moreover, many of the landowners, past whose land the water is carried, tap the conduit lines, with the result that public aqueducts stop their courses for private individuals. . . [31]

76. Concerning wrongs of this sort, however, nothing more or better can be said than that spoken by Caelius Rufus in his public speech entitled *On the Waters:* (2) "Would that we were not proving, by provoking indignant reactions, that all these wrongs are being habitually practiced with comparable impunity: we find fields being watered, taverns, even garrets, and lastly, all the brothels equipped with constantly flowing taps."[32] (3) Now, as to the fact that water of some aqueducts is being distributed in place of others, under false names, even if these are among more trivial shortcomings, yet they must be counted among things that seemed to demand correction.[33] (4) This generally happened around the Caelian and Aventine. (5) These hills, before the Claudia was introduced, used to be supplied by the Marcia and Julia. (6) But after the emperor Nero brought the Aqua Claudia, which he tapped near Spes Vetus, on an arcade to the Temple of the Divine Claudius, so that it could be distributed from that point, the earlier lines were not enlarged but ceased to function. For Nero added no *castella* but used the same ones that had functioned earlier; although their water was changed, their old names remained.

77. Enough now has been said about the capacity of each line and, so to speak, our particular new way of increasing it, as well as the graft and wrongful actions concerning them. (2) It remains for us to discuss in detail the distribution that we found treated collectively and, as it were, in a lump, and indeed even set forth under false identifications. Our discussion will be organized by the individual lines, how each is arranged, and by the regions of the city. (3) Such a treatment, I know, can seem not only boring but also complicated. (4) We

---

31. The text is confused but Frontinus seems to be stressing the contrast between private owners and public needs.

32. Cf. R.H. Rodgers, "Frontinus *Aq*. 76.2: An Unnoticed Fragment of Caelius Rufus?" *AJP* 103 (1982): 333–37.

33. Cf. M.D. Reeve, "Back to the Source," *Liverpool Classical Monthly* 6.5 (1981): 141.

will set it forth, however, as concisely as possible so that nothing may be lacking to what we might call our rule and guide of office. (5) Those satisfied with a knowledge of the totals may skip the detailed discussion.

78. The distribution of 14,018 *quinariae* is carried out in such a way that 771 *quinariae*, which are given by certain lines as a supplement for others and occur twice in the distribution scheme, appear only once in the calculations given. (2) Of these, 4,063 are distributed outside the city:

in Caesar's name	1,718 *quinariae*
to private consumers	2,345 *quinariae*

(3) The remaining 9,955½ *quinariae* used to be distributed within the city through 247 *castella:*[34]

in Caesar's name	1,707½ *quinariae*	
to private consumers	3,847 *quinariae*	
for public functions:	4,401 *quinariae:*	
to 18 (?) camps		279 *quinariae* (379?)
to 95 (?) public works		2,301 *quinariae*
to 39 fountains		386 *quinariae*
to 591 basins		1,335 *quinariae*

But this distribution must also be subdivided according to the names of the separate aqueducts and the regions of the city.

79. Of the 14,018 *quinariae*, then, that we have set down as the total of the distribution of all the aqueducts,[35] only 5 *quinariae* are supplied in the name of the Appia outside the city, since . . . too low. . . .[36] (2) The remaining 699 *quinariae* were distributed within the city through 20 *castella* to Regions II, VIII, IX, XI, XII, XIII, and XIV:

34. Cf. O.A.W. Dilke, review of C. Kunderewicz, *Frontinus, De Aquaeductu Urbis Romae*, *JRS* 67 (1977): 248: "On numerals, in passages where the MSS. vary or are corrupt, the last word has certainly not been said." Here the MSS. give a total of 9,955 *quinariae* distributed inside the city; the subtotals in the MSS. equal a sum of 9,955½. Moreover, the MSS. subtotals for camps, public works, fountains, and basins add up to 4,301 *quinariae*, not 4,401.

35. Cf. Rodgers, "Textual Temptations," 135.

36. Here again the MSS. are confused, with no convincing emendation suggested. Rodgers notes that the sense should be similar to Frontinus' explanation of the Appia's level at 65.7.

in Caesar's name	151 *quinariae*
to private consumers	194 *quinariae*
for public functions:	354 *quinariae:*
to 1 camp	4 *quinariae*[37]
to 14 public works	123 *quinariae*
to 1 fountain	2 *quinariae*
to 92 basins	226 *quinariae*

80. From the Anio Vetus there used to be distributed outside the city 169 *quinariae* in Caesar's name and 404 to private consumers. (2) The remaining 1508½ *quinariae* were supplied inside the city through 35 *castella* to Regions I, III, IV, V, VI, VII, VIII, IX, XII, and XIV:

in Caesar's name	64½ *quinariae*
to private consumers	490 *quinariae*
for public functions:	552 *quinariae:*
to 1 camp	50 *quinariae*
to 19 public works	196 *quinariae*
to 9 fountains	88 *quinariae*
to 94 basins	218 *quinariae*

81. From the Marcia there used to be distributed outside the city 261½ *quinariae* in Caesar's name. (2) The remaining 1,472 *quinariae* were distributed within the city through 51 *castella* to Regions I, III, IV, V, VI, VII, VIII, IX, X, and XIV:

in Caesar's name	116 *quinariae*
to private consumers	543 *quinariae*
for public functions:	491 *quinariae:*[38]
to 4 camps	42½ *quinariae*
to 15 public works	41 *quinariae*
to 12 fountains	104 *quinariae*
to 113 basins	256 *quinariae*

---

37. Here too the subtotals given in the MSS. are inconsistent, adding up to 355 *quinariae*, not 354. This translation here and elsewhere gives the figures of Kunderewicz' text, despite obvious corruptions and inconsistencies.

38. The MSS. omit figures for public functions, with editors generally following those supplied by Poleni. This is also the case for distribution to *privati* by the Aqua Julia at 83.2.

82. From the Tepula there were delivered outside the city 58 *quinariae* in Caesar's name and 56 to private consumers. (2) The remaining 331 *quinariae* were distributed within the city through 14 *castella* to Regions IV, V, VI, and VII:

in Caesar's name	34 *quinariae*	
to private consumers	237 *quinariae*	
for public functions:	50 *quinariae:*	
to 1 camp		12 *quinariae*
to 3 public works		7 *quinariae*
to 13 basins		32 *quinariae*

83. The Julia used to flow outside the city with 85 *quinariae* in Caesar's name and 121 to private consumers. (2) The remaining 548 *quinariae* were distributed within the city through 17 *castella* to Regions II, III, V, VI, VIII, X, and XII:

in Caesar's name	18 *quinariae*	
to private consumers	196 *quinariae* (?)	
for public functions:	383 *quinariae:*	
to 2 (?) camps		69 *quinariae*
to 11 (?) public works		181 *quinariae*
to 3 (?) fountains		67 *quinariae*
to 28 basins		65 *quinariae*

84. In the name of the Virgo there were discharged outside of the city 200 *quinariae*. (2) The remaining 2,304 *quinariae* were distributed within the city through 18 *castella* to Regions VII, IX, and XIV:

in Caesar's name	509 *quinariae*	
to private consumers	338 *quinariae*	
for public functions:	1,457 *quinariae:*[39]	
to 2 fountains		26 *quinariae*
to 25 basins		51 *quinariae*
to 16 public works		1,380 *quinariae*

39. The figures are confused here. I follow the emendation of Buecheler, adopted by Kunderewicz.

(including 460 specifically for the Euripus, to which the Aqua Virgo has given its name).

85. The Alsietina supplies 392 *quinariae*. All of this amount is used outside the city:

in Caesar's name	254 *quinariae*
to private consumers	138 *quinariae*

86. The Claudia and Anio Novus used to be distributed separately outside the city, each from its own channel, but were mixed for delivery inside. (2) The Claudia in fact used to distribute outside the city 246 *quinariae* in Caesar's name and 439 to private consumers; the Anio Novus supplied 728 *quinariae* in Caesar's name. (3) The remaining 3,498 *quinariae* of both lines were distributed within Rome to the 14 regions of the city through 92 *castella:*

in Caesar's name	816 *quinariae*	
to private consumers	1,567 *quinariae*	
for public functions:	1,115 *quinariae:*	
to 9 camps		149 *quinariae*
to 18 public works		374 *quinariae*
to 12 fountains		107 *quinariae*
to 226 basins		485 *quinariae*

87. This supply of water as calculated used to be distributed in this manner up to the time of the Emperor Nerva. (2) Now, thanks to the foresight of our most industrious emperor, whatever was either unlawfully tapped by the watermen or lost through neglect has been added to the total supply, as if by a new discovery of sources. (3) The total capacity indeed has been nearly doubled and subsequently delivered through so careful a distribution system that more than one aqueduct is supplied to regions previously served by only one line, such as the Caelian and Aventine, to which the Aqua Claudia alone used to be brought through the Neronian arches. As a result, whenever some repair had interrupted the supply, these densely inhabited hills lacked water. (4) Now more than one aqueduct supplies them, and in particular the Aqua Marcia, which has been brought back on a large-scale project from Spes Vetus all the way to the Aventine. (5) Moreover, in every part of the city many basins, new as well as old, have received double taps from different lines, so that if an accident may interrupt one of them, service is not stopped, since the other continues its supply.

88. The queen and mistress of the world, who stands as a "goddess on

earth, to whom nothing is equal and nothing comes close,"[40] feels day by day this concern of her most devoted Emperor Nerva. The health and well-being of the eternal city will be affected by it even more, now that the number of *castella*, conduits, fountains, and basins has been increased. (2) No less an advantage is supplied to private consumers from the increase of the emperor's benefits; even those who in fear used to tap water illegally now enjoy it without worry as a result of his grants. (3) Not even waste water is lost; the causes of contagion have been removed, the appearance of the streets is clean, the air is more wholesome, and that atmosphere within our city which was always notorious in the past has been removed.[41] (4) It does not escape me that an account of the new distribution ought to be given here. But it must be understood that things should not be set forth unless they have been completed, when we will have added them to increase the entire supply.

89. What of the fact that not even this is enough for the concern of our emperor? Exhibiting the most painstaking care for his citizens, he believes that he has not contributed enough to our protection and our amenities unless he also makes the water supply that he has added in such great quantity purer and more wholesome. (2) It is worth the effort to review in detail what he has done to enhance the usefulness of all the lines by correcting the shortcomings of particular ones. (3) For when did our city not have turbid and muddy water when even moderate rains had come upon us? (4) This is not because all the lines are like this from their source or because those tapped from springs ought to suffer this injury. This is especially true of the Marcia, the Claudia, and the other lines, whose clarity, which is pure at their intake, is polluted not at all or very little by rainwater, if shafts have been built up along their course and covers set over them.

90. The two Anios continue to be less clear, for they are drawn from the river and often become turbid even in fair weather, since the Anio, although flowing from a very pure lake, nevertheless by its movement carries along from its crumbling banks something that muddies it before it reaches its conduits.[42] (2) It suffers this contamination not only during winter and spring rains but also during those of the summer, at a time when a more pleasing purity is demanded from the aqueducts.

91. One of these lines, to be sure, that is, the Anio Vetus, keeps its pollution to itself, since it is lower in elevation than most others. (2) The Anio Novus,

---

40. Frontinus appears to be quoting Mart. 12.8.1–2.

41. The text is problematic here. The translation follows the revision of Krohn's 1922 Teubner edition, adopted by Kunderewicz.

42. Rodgers, "Textual Temptations," 135.

however, used to contaminate the others; since it comes at a very high elevation and in particularly large supply, it supplements the shortages in other lines. (3) Indeed the incompetence of the watermen, who diverted it into the conduits of other lines more often than was necessary to supplement them, used to contaminate even aqueducts that were of sufficient volume, most of all the Claudia which, although carried for many miles in its own conduit, finally at Rome used to lose, up to this present time, its own high quality after being mixed with the Anio. (4) The Anio Novus in fact did so little to help the waters of the other lines coming in contact with it that the reputations of most of them suffered from the carelessness of watermen who did not distribute the water the way they should have. (5) We have found the Marcia itself, water most pleasing for its purity and coldness, supplying baths and fullers and other needs too foul to mention.

92. It was therefore decided that all the waters of different aqueducts be separated and then be arranged individually so that the Marcia in particular might be used exclusively for drinking, and subsequently that the other lines be assigned suitable functions, each according to its own quality. As a result, the Anio Vetus, less wholesome for several reasons because it is drawn at a lower level, is used for watering gardens and the baser needs of the city itself.

93. Nor has it been enough for our emperor to have restored the capacity and quality of the other lines; he has also recognized that the contamination of the Anio Novus can be prevented. (2) For he has ordered that its water not be drawn from the river but from a lake above the Neronian villa at Subiaco, where it is very pure. (3) Since the Anio River rises above Treba Augusta, whether because it runs through rocky mountains with very few cultivated areas lying around the town itself, or because it loses the sediment from the depth of the lakes into which it is received, being shaded as well by the thickness of the groves which overhang it, it reaches the lake very cold as well as very pure. (4) This water, of such an auspicious quality, the equal of the Marcia in all respects but about to surpass it in capacity, will take the place of that foul and turbid supply, its title indicating in advance its new founder, the emperor Caesar Nerva Augustus.[43]

94. It follows that we should mention the laws concerning the introduction and safeguarding of water, one of which pertains to restricting private consumers to the limits of the grant they have legally obtained, the other to the protection of the conduits themselves. (2) In this, while I was reviewing laws passed

43. The MSS. read "Caesar Nerva Trajan Augustus." Rodgers notes that although Trajan must have completed Nerva's improvements to the Anio Novus, the *auctor* of the project was Nerva, not his successor.

concerning individual aqueducts, I found that certain practices were followed differently among our ancestors. (3) In early times all water was distributed for public functions, and it was so prescribed "that no private citizen draw any other water than that which has fallen from the basin to the ground"—these are the words of that law, that is, water that has overflowed from a basin; we call it "overflow" water. (4) This water in fact used to be allotted for no other purpose than use by baths and fullers' establishments, and was taxed for a set fee to be paid into the public treasury. (5) Some amount also used to be allotted the homes of leading citizens of the state when the others permitted.

95. To what magistracy the right of allotting or selling water belonged is set forth in different ways in the laws themselves. (2) I find that sometimes it was granted by the aediles, sometimes by the censors; but it is evident that whenever there were censors in the republic, the right was primarily sought from them, and when, however, there were not, that power belonged to the aediles.[44] (3) From this it is obvious how much more important a concern our ancestors had for common benefits than for private pleasures, since even water that private consumers drew was related to public needs.[45]

96. I find, however, that contracts for the safeguarding of the individual lines were customarily let out and the requirement was placed on contractors of having a fixed number of artisan slaves around the conduits outside the city and a fixed number inside it. In addition, the requirement was such that they entered on the public records the names of those they would employ in this duty throughout each of the regions. The task of approving such work rested sometimes with the censors and aediles. Sometimes that duty also fell to the quaestors, as is evident from a senatorial decree made during the consulship of Gaius Licinius and Quintus Fabius [116 B.C.].

97. How great was the concern that no one dare to injure the conduits or to tap water not lawfully granted can be shown from many particulars, and especially from this fact: the Circus Maximus was not watered even on days of racing games except by permission of the aediles or censors. (2) We read in Ateius Capito that this practice continued even after the responsibility for the water supply passed to curators under Augustus. (3) Moreover, property which, contrary to law, had been irrigated with public water used to be confiscated. (4) A fine also used to be imposed on the contractor if it was clear that someone had acted without his knowledge against the law.[46] (5) In the same laws the

44. Cf. Rodgers, "Textual Temptations," 135.

45. The translation follows Rodgers' text here; Kunderewicz adopts Buecheler's transposition of 95.3 to a position after 94.4.

46. The translation follows Buecheler's emendation *clam eo* for the *cum eo* of the MSS.

following provision was added: "Let no one with malice contaminate water where it flows for public use. (6) If someone has contaminated it, let there be a fine of 10,000 sesterces."[47] (8) For this reason the curule aediles used to be ordered to appoint two men from those who lived in an individual ward or had property there, under whose judgment water might flow for public use.

98. After his aedileship, which he had carried out after having served as consul, Marcus Agrippa was the first permanent commissioner, so to speak, of his own works and public donations. (2) He set forth at that time, according to the supply available, what water should be allotted to public works, what to basins, and what to private consumers. (3) Agrippa had as well his own company of slaves for the aqueducts to guard the conduits, *castella,* and basins.

99. This company Augustus gave to the state after it had been left to him by Agrippa in his inheritance. (2) After him, in the consulship of Quintus Aelius Tubero and Paulus Fabius Maximus [11 B.C.], decrees of the senate were passed and a law was published about this matter, which up until that time had lacked a fixed law because it had been managed, as it were, by personal authority. (3) Augustus also defined in an edict the right those should enjoy who had concession of water according to Agrippa's records, since the whole system of distribution was now transferred to the emperor's grants. (4) He also established the units of measurement already discussed and appointed Messalla Corvinus as commissioner for maintaining and running the system; Postumius Sulpicius, a former praetor, and a junior senator Lucius Cominius were also made his assistants. (5) Regalia was granted to them as if public officials, and a senatorial decree was passed concerning their office, which is given below:

> 100. Whereas the consuls Quintus Aelius Tubero and Paulus Fabius Maximus have inquired of the senate what it pleases to order concerning the honors of those who have been named commissioners of the public aqueducts by Caesar Augustus with its consent, be it resolved that those who are in charge of the public aqueducts, when they are outside the city for the sake of this business, have two lictors and three public slaves, individual architects, secretaries and clerks, and as many attendants and heralds as those officials through whom grain is distributed to the people. (2) When, however, they carry out some official business in the city, they should have the same attendants except the lictors. (3) Be it resolved that they report to the treasury within ten days of this decree those attendants whom they, as commissioners of the aqueducts, are permitted to use, and that to those who have been reported they

47. Chapter 97.7 is deleted by most editors as a later gloss.

> give and allot, as wages from the praetor of the treasury, as much annual provisions as the prefects of the grain supply are accustomed to give and report, and that they be permitted to use those moneys with impunity. (4) Moreover, the consuls Quintus Aelius and Paulus Fabius, both or one of them, if it will seem best, should contract for supply of tablets, paper, and other required materials to be furnished those commissioners for discharge of their office, through the praetors in charge of the treasury as intermediaries.
>
> 101. Likewise, since the commissioners of the roads and grain supply perform their public duty during a fourth part of the year, be it resolved that the commissioners of the aqueducts be exempt from judicial duties private and public for the same period of time.

(2) Although the treasury persists in paying for these attendants and services, they seem to have lost their effectiveness because of the laziness and inactivity of the commissioners who do not carry out their office. (3) Moreover, the senate had ordered that lictors be present for commissioners leaving the city on official business. (4) For us, as we make the rounds of the conduits, our own good faith and the authority granted us by the emperor will take the place of the lictors.

102. Since we have brought our account to the establishment of the commissioners, it is not out of place to add the names of those who have been in charge of this office, after Messalla, up to our own time.[48] (2) Ateius Capito succeeded Messalla in the consulship of Plancus and Silius [A.D. 13]. (3) Tarius Rufus succeeded Capito in the consulship of Gaius Asinius Pollio and Gaius Antistius Vetus [A.D. 23]. (4) Marcus Cocceius Nerva, the grandfather of the Divine Nerva, also noted for his knowledge of the law, succeeded Tarius in the consulship of Servius Cornelius Cethegus and Lucius Visellius Varro [A.D. 24]. (5) Gaius Octavius Laenas succeeded him in the consulship of Fabius Persicus and Lucius Vitellius [A.D. 34]. (6) Marcus Porcius Cato succeeded Laenas in the consulship of Aquila Julianus and Nonius Asprenas [A.D. 38]. (7) . . . succeeded him after . . . in the consulship of Servius Asinius Celer and Sextus Nonius Quintilianus [A.D. 38]. . . . Aulus Didius Gallus succeeded in the consulship of. . . . (8) Gnaeus Domitius Afer succeeded Gallus in the consulship of Quintus Veranius and Pompeius Longus [A.D. 49]. (9) Lucius Piso succeeded

48. For a detailed commentary on this chapter see R.H. Rodgers, "*Curatores Aquarum*," *HSCP* 86 (1982): 171–80; and Bruun, 153–68. There is almost certainly a gap in 102.7, and there may well be missing names and dates after 102.1 and in 102.16. In 102.13 the curator is Vibius Crispus, not the Albius of the MS. tradition.

Afer in the fourth consulship of Nero Claudius Caesar with Cossus, the son of Cossus [A.D. 60]. (10) Petronius Turpilianus succeeded Piso in the consulship of Verginius Rufus and Memmius Regulus [A.D. 63]. (11) Publius Marius succeeded Turpilianus in the consulship of Crassus Frux and Laecanius Bassus [A.D. 64]. (12) Fonteius Agrippa succeeded Marius in the consulship of Luccius Telesinus and Suetonius Paulinus [A.D. 66]. (13) Vibius Crispus succeeded Agrippa in the consulship of Silius and Galerius Trachalus [A.D. 68]. (14) Pompeius Silvanus succeeded Crispus in the third consulship of Vespasian with Cocceius Nerva [A.D. 71]. (15) Tampius Flavianus succeeded Silvanus in the second consulship of Domitian with Valerius Messallinus [A.D. 73]. (16) Acilius Aviola succeeded Flavianus in the fifth consulship of Vespasian, which was the third consulship of Titus [A.D. 74]. (17) After him, in the third consulship of the Emperor Nerva and Verginius Rufus [A.D. 97], responsibility for the office was transferred to me.

103. I will now add those things that the commissioner of the water supply ought to observe and the law and decrees of the senate applicable to determining his conduct of office. (2) Regarding the law on drawing water by private consumers, it must be observed that no one tap water without written authorization from Caesar, that is, that no one tap water not granted, and further, that he not tap more than he has legally been granted. (3) In this way, then, we will insure that the additional volume, which we said has been acquired, can be used for new fountains and new grants of the emperor. (4) In each case, however, great care must be taken to counter all sorts of dishonesty. The conduits must be repeatedly checked with care outside the city to verify the authorized grants, and the same must be done in the *castella* and public fountains, so that the water may flow without interruption day and night. (5) The commissioner is also ordered to do this by decree of the senate, the words of which are as follows:

> 104. Whereas the consuls Quintus Aelius Tubero and Paulus Fabius Maximus have inquired of the senate what it pleases to order concerning the number of public fountains Marcus Agrippa built in the city and within buildings adjoining the city, be it resolved that the number of public fountains, which those to whom the senate has given the task of inspecting the public aqueducts and calculating the number of public fountains have reported to be in existence, be neither increased nor decreased. (2) Likewise be it resolved that the water commissioners, whom Augustus Caesar has named by authority of the senate, give attention that the public fountains deliver water as constantly as possible day and night for the needs of the people.

(3) In this senatorial decree I believe it should be noted that the senate forbade the number of public fountains to be increased as well as decreased. (4) I think this was done because the volume of the aqueducts that were coming into the city in those times, before the Aqua Claudia and Anio Novus were introduced, did not seem adequate to support a greater distribution.

105. Anyone who wishes to draw public water for his own personal needs will have to obtain it by grant and bring written authorization from the emperor to the commissioner. The commissioner must expedite Caesar's grant and immediately send notice in writing to the freedman of Caesar, imperial deputy of the same office. (2) Tiberius Claudius, moreover, seems to have been the first to have added such a deputy after he introduced the Anio Novus and the Claudia. (3) The letter's contents must also be made known to the superintendents, so that they may not excuse their negligence or dishonesty on the grounds of ignorance. (4) After summoning the surveyors, the deputy should take care that the pipe of the unit of measurement granted be so marked; he should give special attention to the capacity of the measures we have indicated above and should make note of its positioning so that it not be left to the discretion of the surveyors to approve a pipe of sometimes larger or sometimes smaller capacity for the benefit of the parties involved. (5) The deputy must not allow free judgment in adjoining any type of lead pipe, but one of the same size as that indicated on the connecting pipe for the length of 50 feet, as is provided by the following senatorial decree:

> 106. Whereas the consuls Quintus Aelius Tubero and Paulus Fabius Maximus have inquired of the senate what it pleases to order in the case of certain private consumers drawing water from public conduits, be it resolved that no private citizen be permitted to tap public conduits and that all, to whom the right of drawing water has been granted, draw it from *castella,* and that the water commissioners consider in what places inside and outside the city such consumers can suitably build *castella* from which they may draw public water which they have received from a *castellum* through the water commissioners. (2) He to whom public water is granted should not have the right to adjoin a pipe larger than a *quinaria* . . . within 50 feet of the *castellum* from which they draw the water.[49]

(3) In this senatorial decree it is worthy of note that it does not permit the

49. The MS. tradition is corrupt here, since this statement as given does not agree with that in 105.5. Cf. Bruun, 111; and Pace, 301.

drawing of water except from a *castellum,* so that neither the conduits nor the public pipes be frequently tapped.

107. The right of drawing water passes neither to an heir nor a buyer nor any new master of a property. (2) In times past the privilege used to be granted to baths open to the public that water once granted should remain so continually. (3) We learn this from old senatorial decrees, one of which I have added below. Now every grant of water is renewed with a change of owner.

> 108. Whereas the consuls Quintus Aelius Tubero and Paulus Fabius Maximus have informed the senate that it is necessary to determine by what law those to whom water has been granted may draw it inside and outside the city and have asked what it pleases to order concerning this matter, be it resolved that a grant of water remain as long as the same owners hold that land onto which they have drawn water, with the exception of water granted for the use of baths or under the right of drawing.

109. When water from aqueducts begins to go unused, an announcement is made and an entry is made in the record books, which are reviewed so that unused water may be granted to those having applied for it. (2) Previously they used to cut off such supplies at once so that they could sell them in the meantime, either to those holding the property or even to other individuals. (3) It has seemed more generous to our emperor that such supplies to properties not be interrupted immediately but that a period of 30 days be granted, within which those concerned with such matters may make formal application. (4) Concerning water granted to the property of partners, I find no fixed rule. (5) Nevertheless, it is the practice, as if prescribed by law, that as long as one of those who have collectively been granted water survives, the total grant as assigned to the property should continue to be furnished and the grant should finally be renewed at that time when each of the parties to whom it was given has ceased to possess the property. (6) That water be brought to a place other than that property to which it will have been granted, or be drawn from a *castellum* other than that from which it will belong by written authorization of the emperor, is not only manifestly not allowed but also prohibited by ordinances.

110. Water that is designated "overflow," that is, that which is the overflow from *castella* or the discharges of pipes, is also legally obtained. (2) This grant is usually made most sparingly by the emperors, but it is subject to the dishonest practices of the watermen. How much concern is needed for preventing such things will be evident from the heading of an ordinance I have added below:

111. "I wish no one to draw 'overflow' water except those who obtain it by

my own grant or that of earlier emperors. (2) For it is necessary that some part of the water overflow from the *castella*, since this concerns not only the wholesomeness of our city but also the advantage of flushing out the sewers."

112. Since we have set forth those things that used to concern the regulation of water for private use, it is not inappropriate to touch on certain examples of actions by which we in simple performance of duty discovered the most wholesome of arrangements being thwarted. (2) I have found in many *castella* certain delivery fittings larger than the amounts that had been granted, and some of these not even stamped with their capacity. (3) Whenever a stamped fitting exceeds its lawful capacity, dishonesty is detected on the part of the deputy who stamped it. (4) When, however, it is not even stamped, the obvious fault of all is discovered, most of all that of the party being supplied, then that of the overseer. (5) In certain *castella*, although delivery fittings of legal capacity had been stamped, pipes of larger volume were joined to them, with the result that the water was not carried through an opening of legal size but, because it was forced through a small narrow pipe, it readily filled the larger pipe connected to it. (6) Care must also be taken that whenever the capacity of a fitting is stamped, the pipes to which it leads must also be stamped for the distance prescribed by the senatorial decree cited above. (7) In this way then, the overseer will lack any excuse for misconduct, since he has understood that only pipes with stamped capacity ought to be connected.

113. In the joining of delivery fittings, care must be taken that they be installed at the same level and that one person's fitting not be installed too low or a second person's too high. (2) A lower fitting draws more, a higher one less, since the current of the water is taken by the lower one. (3) In the fittings of certain *castella*, connections have not even been installed. (4) These are called "loose" and are enlarged or narrowed as it has pleased the watermen.

114. Moreover, the following dishonest practice of the watermen is not to be tolerated: when water rights have been transferred to a new owner, they install a new outlet in the *castellum* but leave the old one through which they draw water, which they sell. (2) I think therefore that this practice in particular must be corrected by the water commissioner. (3) Indeed it concerns not only the safeguarding of the water itself but also the protection of the *castellum*, which is damaged when tapped repeatedly and without reason.

115. Likewise the following source of revenue for the watermen must be done away with, that which they call "punctures." (2) There are all over the city different areas of some distance through which pipes run concealed under the pavement. (3) I have found that these pipes, which have been tapped everywhere by one who used to be called "the puncturer," supplied water to all the businesses en route by private branches; as a result, a limited amount was

delivered for public needs. (4) How much water of this supply has been reclaimed I judge from the fact that a considerable quantity of lead has been collected when branches of this sort have been removed.

116. There remains the safeguarding of the conduits. Before I begin to discuss this topic, a few matters should be explained about the team of slaves established for this purpose. (2) There are two teams belonging to the water commission, one public, the other the emperor's. (3) The older team is the public one, which we said was left by Agrippa to Augustus, who turned it over to the state; it has about 240 men. (4) Caesar's team, which Claudius established when he brought his aqueducts into the city, is 460 in number.

117. Each team, moreover, is divided according to several classifications of duties: overseers, *castellum* men, inspectors, pavers, plasterers, and other artisans. (2) It is necessary that some of these be outside the city for things that are not of major construction but still seem to demand prompt attention. (3) The men in the city around the stations of the *castella* and ornamental fountains will apply themselves to all necessary tasks, especially in the event of unexpected misfortunes, so that the protection of surplus water can be directed from other districts to that upon which an emergency has fallen. (4) It has become customary, either because of bribery or the neglect of their overseers, that a considerable number of each team is diverted to private tasks. We have decided to call them back to their public duties and some discipline in such a way that we may direct what they do one day before and enter in the records what they have accomplished each day.

118. The costs of the public team are paid from the treasury, an expense that is lightened by revenue derived from taxes related to water rights. (2) These are received from the places or buildings that are around the conduits, *castella,* ornamental fountains, or basins. (3) This revenue of nearly 250,000 sesterces, previously lost and unaccounted for, indeed in most recent times paid into the private coffers of Domitian, the justice of the Divine Nerva has restored to the people. We have taken care to bring it under a fixed rule so that places subject to this water tax may be established and well known. (4) Caesar's team takes its wages from the imperial treasury, from which also the cost of all the lead and all expenses pertaining to the conduits, *castella,* and basins are paid.

119. Since we have set forth those things that seemed to relate to the teams, we will turn as promised to the maintenance of the conduits, a matter of more earnest concern, since they are a most eminent testimony to the greatness of the Roman Empire. (2) Many extensive tasks continually arise that ought to be attended to before they begin to require major assistance. Nevertheless, they must generally be undertaken with wise restraint, since one ought not always believe those seeking to build or extend an aqueduct. (3) For this reason, the

commissioner should be supplied not only with the knowledge of professionals but also with his own practical experience. He should use not only the architects of his own office but should call upon the loyalty, no less than the careful knowledge, of others, in order to judge what things should be done without delay, what should be postponed, what should be carried out by contractors, and what by artisans directly under his control.

120. Tasks arise from these causes: damage occurs either from age, the unlawful behavior of nearby owners, the force of the elements, or faultiness of construction, a defect that has happened more frequently in newer conduits.

121. Generally it is those sections of conduit that are supported by arcades or attached to the sides of hills that are affected by age or the force of the elements, and of those on arcades, the sections that are carried across a stream. (2) This sort of repair therefore should be undertaken with careful dispatch. (3) Underground sections are less subject to damage, since they are not exposed to extremes of cold or heat. (4) Defects, moreover, are of the following sorts: either those that are corrected without interruption of the supply or those that cannot be corrected except by diverting the flow, such as repairs that must be carried out in the channel itself.

122. This last type of repair arises from a twofold cause: either the course of the water is narrowed by accumulation of deposit, which sometimes hardens into a crust, or the interior lining is damaged; the result is leaks, which of necessity weaken the sides of the conduits and the substructures. (2) Supports themselves that are constructed of tufa give way as well under such great weight. (3) Repairs around the channels of the conduits ought not to be made in summer, so that supply not be interrupted at a time when it is especially needed, but in spring or autumn, and with the greatest dispatch so that, when all arrangements have, of course, been made in advance, the flow in the conduits should be halted for as few days as possible. It escapes no one that this sort of repair must be done on one aqueduct at a time, so that water not be lacking to the city if more than one were interrupted in the same way.

123. Those repairs that ought to be carried out without interrupting the flow of water consist most of all of structural work, which must be done in a reliable fashion and at the proper times. (2) The period suitable for such construction is from April 1 through November 1, recognizing that it is best to avoid that part of the summer which glows with excessive heat, since there is need for moderate weather, so that the concrete when laid may properly absorb its mortar and be hardened into a single mass. Moreover, an extreme cold damages concrete no less than the sun. (3) Nor does any construction demand more diligent care than that expected to withstand the force of moving water: in such work, therefore, reliability must be demanded in individual details, according to a rule known to all but heeded by few.

124. I would believe that the following is doubtful to no one: the conduits that are closest to the city, that is, those that are built of squared stone inside the sixth milestone, must be safeguarded to the greatest extent, since they are of the largest construction and their individual arcades carry more than one aqueduct. (2) If it is necessary to shut them off, the city will be deprived of the greater part of its water.[50] (3) There are, however, remedies even for difficulties of this kind: a temporary channel is raised at the level of the conduit not in use, and the channel is extended by lead troughs connected to it for the length of the interruption. (4) Moreover, since almost all the lines had been laid out through territory under private ownership, and preparation for future expenditure seemed difficult were it not aided by some provision of the law, and that contractors also not be prevented by landholders from repairing conduits, the following decree of the senate was passed:

> 125. Whereas the consuls Quintus Aelius Tubero and Paulus Fabius Maximus have spoken about repair of the conduits, channels, and arches of the Aqua Julia, Marcia, Appia, Tepula, and Anio and have inquired of the senate what it pleases to order on the subject, be it resolved that when those conduits, channels, and arches are repaired, which Augustus Caesar has promised the senate he will repair at his own expense, earth, mud, stone, potsherds, sand, wood, and other things necessary for this task, be given, removed, taken, and carried out from private property in the nearest vicinity where each of them can be removed, taken and carried without injury to their owners after their value has been established by the judgment of a good man; and for the removal of all these things for the purpose of rebuilding these structures, as often as there is need, that roads be given and rights of passage be open through the property of private owners without injury to them.

126. Frequently, however, damages arise from the lawlessness of proprietors who injure the conduits in many ways. (2) Indeed they first occupy with buildings or trees the areas around the aqueducts which, according to senatorial decree, ought to be free. (3) Trees do more harm, since the vaulting and the sides of the channels are broken by their roots. (4) Then proprietors lay out neighborhood and country roads across the aqueducts themselves. (5) Last of all, they prevent access for their maintenance. (6) All of these things have been provided for in the following senatorial decree:

> 127. Whereas the consuls Quintus Aelius Tubero and Paulus Fabius

50. Cf. Rodgers, "Textual Temptations," 135–36.

Maximus have informed the senate that the rights of way of aqueducts that come into the city are being occupied by tombs and buildings and planted with trees, and have inquired what it pleases to order on this subject, be it resolved that, . . . through even which public works are being damaged,[51] it is decreed that a space of 15 feet be kept free on each side around sources, arches, and walls, and that around the conduits that are underground and channels inside the city and buildings adjoining the city five feet be left free on each side, in such a way that it not be permitted after this time to put a tomb or a building in these places or to plant trees; if now there should be any trees within this space, that they be cut out unless they are part of a farm and enclosed by structures. (2) If someone shall have acted against these restrictions, the fine shall be 10,000 sesterces for each offense, half part of which shall be given as a reward to the informer by whose effort he who acted against this senatorial decree has been chiefly convicted; half part, however, shall be paid to the treasury. (3) The water commissioners shall also judge and make inquiry into this matter.

128. This senatorial decree could seem most just, even if those spaces were claimed only from the standpoint of public benefit. But it is much more so, since our ancestors with an admirable sense of justice did not confiscate from private owners even those lands that related to the public good. When they introduced aqueducts, if a proprietor had been a bit difficult in selling part of his land, they offered money for all the property and after the requisite sections had been fixed, they sold back to him the excess property, so that the state as well as private citizens might have their own rights, each within its proper boundaries.[52] (2) Many people, however, not content to have occupied their own boundaries, have made attacks against the conduits themselves through punctures everywhere in the sides of the channel . . . They hold the legally granted right of water . . . than those who, to however small a degree, on the pretext of gaining this privilege, abuse . . . for attacking the conduits.[53] (3) What further should be done, were not all the practices prohibited by a most careful statute and no light punishment threatened for those stubborn enough to violate it? (4) I have therefore given the words of the law:

---

51. The MSS. are confused here, and some phrase indicating removal of obstructions seems to have been lost.

52. Cf. Rodgers, "Textual Temptations," 136.

53. This is a very problematic passage, for which no convincing explanation can be given.

129. Titus Quinctius Crispinus the consul duly put the question to the people, and the people voted in the forum in front of the rostra of the Temple of the Divine Julius on June 30 [9 B.C.]. (2) The Sergian tribe was first. (3) For this tribe Sextus Vibidius Virro, the son of Lucius, voted first. (4) Whoever after passage of this statute shall intentionally and with malice pierce, break, or cause the piercing or breaking or shall make a shaft into the conduits,[54] channels, arches, pipes, spigots, *castella,* and basins of the public aqueducts that are brought and will be brought into the city, so that those waters or any of them not be able to go, fall, flow, be delivered, and be brought into the city of Rome or that in the city of Rome and in those places and buildings that are or will be adjacent to the city, in those properties, estates or places, to the masters, proprietors, and users of these gardens, estates or places to whom water has been given or will be, so that it not issue, be furnished, be distributed, and be allotted to *castella* and basins, may he be condemned to pay 100,000 sesterces to the Roman people. (5) And he who shall have done anything of these things,[55] let him be condemned to repair, rebuild, restore, build, replace, and swiftly demolish without malice. For the enforcement of all these things by fine, let whoever will be the water commissioner, and if no one will be the water commissioner, then let that praetor who exercises judgment between citizens and foreigners, fix and exact it by fine or bail. And to that commissioner, or, if there will not be a commissioner, then to that praetor for this purpose let there be the right and power of arrest and taking bail. (6) If a slave shall do anything of these things, let his master be condemned to pay 100,000 sesterces to the Roman people. (7) If some place has been enclosed or will be enclosed around the conduits, channels, arches, pipes, spigots, *castella,* and basins of the public aqueducts that are brought and will be brought to the city of Rome, let no one in that place after the passage of this statute put in the way, construct, enclose, fix, set up, place, arrange, plow, sow, or admit anything into it except for the sake of constructing and repairing things, an action that will be permitted and required under this statute. (8) If anyone shall do anything against these things, against him let there be the same statute, the law, and procedure of all things and for all things, just as it is and ought to be if someone has broken or punctured a conduit or channel in violation of this statute. (9) That not to graze herds in this place or

54. Cf. Rodgers, "Textual Temptations," 136.

55. The translation follows Rodgers' text here; Kunderewicz gives *et qui clam.*

to cut grass or hay, brambles . . . the water commissioners, who now are and will be . . . [in that place] which has been bounded around the sources, walls, conduits, and channels, let trees, vines, shrubs, brambles, banks, enclosures, thickets, and beds or reeds be removed, cut out, dug out, and rooted out, as he will wish which thing to have been done rightly. And for this purpose let there be for them the right to take bail, levying of fines, and arrest. And let them be permitted to do this without prejudice, let there be right and power. (10) And that the vines and trees which have been enclosed in farms, buildings or walls, and walls, which the water commissioners on learning the reason for them have permitted the owners not to be demolished, on which the names of the commissioners themselves who have allowed this have been inscribed or engraved, may remain, let nothing be asked under provision of this statute. (11) That from those sources, conduits, channels, and arches of aqueducts that it not be permitted to take or draw water to whomever the water commissioners have given or will give permission, except by wheel, water snail,[56] or mechanical device, provided no well shaft or new tap be made, let nothing of this be asked under provision of this statute.

130. I would not deny that transgressors of this most useful statute are worthy of the punishment that is threatened, but because of negligence over a long period, it has been necessary that those caught be called back gently. (2) We therefore worked carefully, as much as we could, so that those who had erred remained unknown. (3) But for those, who after being warned, sought the emperor's pardon, we can appear to be the reason for the legal grant they have obtained. (4) Against the others,[57] however, it is my desire that enforcement of the law not be necessary, since it is more important to safeguard the good faith of my office even if it means giving offense.

56. Cf. Rodgers, "Textual Temptations," 136.
57. Cf. Rodgers, "Textual Temptations," 136.

CHAPTER 2

Frontinus' "Rule and Guide"

There has been ample scholarly discussion of Frontinus' life and career, subjects interesting in themselves but of little direct relevance to the problem of water distribution in the city.[1] The contents and purpose of *De aquaeductu urbis Romae* itself, however, merit more detailed examination. The author states early in his treatise that he has compiled *De aquaeductu* as a *commentarius*[2] or notebook for himself or for a possible successor, to serve as a *formula administrationis*, a "rule and guide" of office (2.2–3; cf. 77.4, *formula officii;* and 37.1, *formulas modulorum*).[3] Pierre Grimal stated the obvious when he observed that Frontinus' words are immediately suspect:[4] if the *commentarius* were primarily for his own edification, why did he go to the trouble of publishing it?

---

1. W. Eck, "Die Gestalt Frontins in ihrer politischen und sozialen Umwelt," in *Wasserversorgung*, 47–62, gives a full account of Frontinus' career. For summaries cf. Hodge, 16–18; Bruun, 10–11; Bennett, xiii–xxvii; Ashby, 26–27; Grimal, vi–viii; Pace, 200–201; and A. Kappelmacher, "Sex. Iulius (Frontinus)," *RE* 10 (1919): 591–95.

2. On the term *commentarius* as applied to *De aquaeductu* see F.R.D. Goodyear, "Technical Writing," in *Cambridge History of Classical Literature*, ed. P.E. Easterling and B.W. Knox, vol. 2, *Latin Literature* (Cambridge, 1983), 672: "A *commentarius* could be a published composition in the plain style, or lack polish altogether: there was no firm tradition, as for the major genres of prose. The subject matter and the author's personality, rather than rules of genre, determined the character and quality of the writing. . . . one finds occasional embellishments and may fairly suppose that Frontinus wrote for a wider public than experts like himself."

3. The word *formula* is itself unusual. For this meaning of "rule, guide, paradigm" see *Thesaurus Linguae Latinae* (Leipzig, 1926) 6.1114, s.v. B 9. Grimal, 65 n. 1, compares Frontinus' phrase *formula administrationis* to the edicts of magistrates, through which he announces principles that will guide his conduct of office.

4. Grimal, xv. Scholars continue, however, to take Frontinus' statement at face value. Cf. Quilici, "Acquedotti," 59; and Robinson, "Water Supply," 44: "he wrote his treatise to inform himself, much as Pepys collected information in his early years at the Admiralty."

There is really no clear parallel to Frontinus' treatise in the body of extant Roman literature. While some scholars have suggested that Vitruvius' book on architecture provides a model,[5] Vitruvius' purpose in setting forth the "principles of modern architecture" was far different from that of Frontinus, a senatorial commissioner describing his duties in office.[6] Most commentators who have addressed the question have followed Ashby in considering Frontinus' treatise a celebration of Trajanic reforms by a dedicated public servant,[7] and therefore a political statement affirming the virtues and benefits of senatorial administration represented by his assumption of duties as *curator aquarum*.[8] Grimal, emphasizing Frontinus' praise of the emperor (1.1, 87.2, 88.1, 89.1), goes so far as to describe the treatise as propaganda celebrating Trajanic policies.[9]

Bruun, who has discussed the problem most recently, is skeptical about a political motivation for the work, but recognizes that Frontinus' omission of details about Flavian administration may reflect a conscious *damnatio memoriae* concerning Domitian's principate under Nerva and the early years of Trajan; he suggests instead that Frontinus' "obvious intention" was to demonstrate to the "educated general reader" the greatness of Rome through its aqueducts.[10] Given this wide range of interpretations, more attention should be focused upon Frontinus' own statement of purpose in light of the book's contents.

Despite its claim to be a *formula administrationis* for its author, *De aquaeductu* is hardly a comprehensive account of the aqueduct system.[11] Frontinus himself gives a summary of its contents in his third chapter, and the overall plan of the work demonstrates considerable selectivity in the topics covered:

1–3	Introduction
4–16	History of Water System
17–22	Elevations of Individual Lines

5. Bruun, 15–16, cites as "more than a remote parallel" Vitruvius' statement in *De arch*. 6 *praef*. 5, *sed tamen his voluminibus editis, ut spero, etiam posteris ero notus*.

6. See F.E. Brown, "Vitruvius and the Liberal Art of Architecture," *Bucknell Review* 11.4 (1965): 99–107; Hodge, 15–16.

7. Pliny *Ep*. 9.19.6 presents Frontinus as a modest, hardworking, and loyal administrator, a portrait echoed by Landels, 211–15. Cf. D'Amato, 182–83.

8. Ashby, 26–27: "this is an aspect of Frontinus which has been overlooked." Cf. Rodgers, "*Copia Aquarum*," 353.

9. Grimal, vi–xvi. Cf. Pace, 202.

10. Bruun, 14–15, 18–19. Elsewhere, however, Bruun presents a somewhat different view, that the book was also intended to immortalize Frontinus himself and his work as *curator* (178–79, 186–87), but he does not explore this idea fully.

11. Grimal, xiv; Bruun, 17–18, 369–70; R.H. Rodgers, "An Administrator's Hydraulics: Frontinus *Aq*. 35–36.2," in *Currents*, 15–19; cf. Hodge, 16–18.

23–63 Pipe Dimensions
64–76 Delivery of Individual Lines
77–86 Distribution Figures for Lines
87–93 Trajanic Improvements
94–130 Administration and Maintenance

Almost one-third of the treatise, 40 chapters from a total of 130, is devoted to the dimensions and capacities of 15 individual pipes in use within the system, and also those of ten pipes not in use (41, 44, 47, 49, 51, 53, 55, 57, 59, 61). This subject is discussed twice, both in theoretical terms (23–36), then again in a detailed catalog of individual pipes (37–63). This may strike the modern reader as odd, but for Frontinus the capacity of the pipes provided the principal basis for determining the quantity of water delivered to the city for subsequent distribution.[12] Scholars have also speculated why he would have bothered to include within his catalog ten pipes not in use in the water system.[13] Certainly this list could have been abridged, or omitted altogether.

Frontinus' treatment is also uneven in other topics: in discussing the problems of administration and maintenance, to which he devotes 35 chapters, he cites the need for regular upkeep (120–24) but omits practical details, such as the mechanics of repair and the costs involved.[14] Instead he quotes at length six senatorial decrees (100–101.1, 104.1–2, 106.1–2, 108, 125, 127), an imperial edict (111), and two laws (97.5–6, 129) on the safeguarding of the lines, taking care to point out to readers the special pertinence of the documents cited (97.1, 99.5, 103.4–5, 105.6, 107.3, 110.2, 124.4, 126, 128).

Bruun has rightly observed, however, that the *senatus consultum* Frontinus cites in 104.1–2 had lost all relevance following the introduction of the Claudian aqueducts, which had dramatically increased water delivery to the city in the first century A.D. We might well ask why Frontinus included it, and whether other *senatus consulta* or laws cited in the treatise might also have been irrelevant or superseded.[15] Certainly our archaeological evidence does not provide

---

12. Rodgers, "*Copia Aquarum*," 353–54; cf. Robinson, "Water Supply," 69–70: "from the long section on adjutages it is clear that inspecting them and pipes which led to private users also fell within the curator's duty of supervision."

13. Grimal, 84, describes the pipes not in use as "*tuyaux theoretiques*"; Bruun, 58, suggests that Frontinus' list is a "desktop-regulation" perhaps found in imperial *commentarii* and included in his treatise for the sake of completeness.

14. Cf. Bruun, 370: "If someone wanted to know exactly how everything in the *cura aquarum* was run, and what the legislation said about such things, then the *De aquae ductu* would not have satisfied him."

15. Bruun, 105–6.

confirmation of many procedures set forth in *De aquaeductu*, such as the stamping of pipe capacities for the first 50 paces of *fistulae* attached to a spigot (112.6–7), not to mention the very use of a *calix* itself, which Frontinus chose to describe in some detail (36).[16] Other features of the treatise are also unusual. Frontinus begins his book with a historical account of the aqueduct system from the first line, the Aqua Appia, to the most recent Claudian aqueducts (4–16); these thirteen chapters are sketchy and statistical, focusing on the starting points, courses, and lengths of individual conduits both above and below ground.[17] Frontinus' lack of attention to historical detail has resulted in scholarly controversies wherever the text is faulty, such as his account of the introduction of the Aqua Marcia to the Capitol (7.4). The author even apologizes for including a history of the lines, claiming a desire not to omit anything (3.1).

The section of the work most valuable for the problem of water distribution—Frontinus' account of the delivery of individual aqueducts within the city—is given in the barest statistical form (*quam brevissimum*, 77.4). Frontinus explains that some readers may find it not only boring but also too complicated (*comprehensionem . . . non ieiunam tantum sed etiam perplexam*), and alerts readers satisfied with a knowledge of the totals that they may skip the detailed discussion (77.3–5). Moreover, the author deviates from his own outline of contents set forth in his introduction (3),[18] including in the treatise two subjects not listed: an extended account of his own measurements of individual lines (64–76), which he indicates are 10,000 *quinariae* greater than the capacities indicated in the imperial record books (74.4), and a report on Trajanic improvements (87–93), which he states he will not describe in detail because they are still incomplete, but, as he declares, will dramatically improve water delivery within the city (88.4). Such discrepancies may be evidence that Frontinus revised an earlier version of his *commentarius* for publication at a later date.[19] Certainly they call into question his presentation of his treatise as a personal guide or one intended for successors (2.2–3).[20]

---

16. Bruun, 39–62, discusses these discrepancies in some detail, pointing out (41 n. 8) that no bronze piece corresponding to a *calix* as described by Frontinus has to date been found; that numerals appearing on *fistulae* stamps frequently indicate data other than capacity (50–52); and that pipes without stamps have been found installed within 50 paces of a *castellum* (61–62). Such discrepancies, he argues, suggest that Frontinus' treatise is often prescriptive rather than descriptive in nature.

17. Grimal, x–xi, suggests that his principal sources for this history were official archives, but Frontinus himself presents evidence that he consulted other sources, particularly for lines introduced before Agrippa's aedileship (7.4–5).

18. See Grimal, ix–x, 65 n. 2; Pace, 237 n. 3.

19. Grimal, x; Rodgers, "*Copia Aquarum*," 353.

20. Cf. Bruun, 369–70: "The book was hardly written for the main purpose of

De aquaeductu is intended for a general readership, making a statement that is at once both public and self-promoting. In it Frontinus does far more than present the aqueduct system as one of the "marvels of the Roman world" in the spirit of Vitruvius and Pliny.[21] Indeed, while Frontinus praises the aqueducts as an *indicium* of Roman greatness (119.1), there is much negative comment in the book. In describing the administration and maintenance of the water system, he repeatedly states that he has inherited a corrupt and inefficient staff (9.6–7, 31.4, 32.6, 34.5, 66.7, 67.8, 72.4, 75.2–3, 109.2, 112.1–4, 113.3–4, 114–15), who must be carefully supervised (117.4). Frontinus indeed seems to discount the professional expertise of his subordinates, describing them somewhat arrogantly as the "hands and tools" of the *curator* (2.1).[22] Earlier commissioners, although described in flattering terms as *principes civitatis* in the opening chapter (1), are later cited as careless in their record keeping and conduct of office (74.2, 101.2–4, 117.4).[23] Frontinus, however, asserts that his own *sollicitudo* and *fides* are modeled on those of the emperor himself (1, 64.1, 130).

To be sure, while the official post of *curator aquarum* had become largely ceremonial by the late first century,[24] the responsibilities of this position were considerable in administering the water supply. Rome's aqueduct system, essential to the life of the city (1, 124.1–2), had been traditionally subject to abuses (7.1). As early as the second century B.C., Cato as censor shut off public water being tapped by private consumers (Livy 39.44.4), and a famous letter of Marcus Caelius Rufus to Cicero cites wrangling with the staff of the water system during his term as curule aedile (*Fam.* 8.6.4).[25] The very size and complexity of the *cura aquarum* made it vulnerable to mismanagement and fraud;[26] the commissioner had a staff of some seven hundred men to supervise (116.3–4).[27] Frontinus, however, appears to give an especially bleak presentation of conditions before

---

providing Frontinus himself with help in his office . . . nor can its content as a whole have been essential for his successors in the post of *curator aquarum*."

21. Bruun, 18, compares Frontinus' presentation to treatments of technical subjects found in journals like *Scientific American*.

22. Frontinus' attitude toward subordinates and his approach to his position are strikingly like those of the Younger Cato in his quaestorship, as presented by Plut. *Cat. Min.* 16–17.1. I owe this parallel to Dr. Nicholas M. Horsfall.

23. Cf. Rodgers, "*Copia Aquarum*," 357 n. 15: "*facere aestimationem* [in 74.2] pointedly contrasts with Frontinus' own *mensurae*."

24. See the discussion of Bruun, 183–87, who describes the office as "paid leave."

25. Cf. Rodgers, "Frontinus *Aq.* 76.2," 333–37. For a survey of Republican administration of the water supply see Robinson, "Water Supply," 50–56.

26. Cf. Coarelli, "*Suburbium*," 43–44, and Champlin, 108–9, for fraudulent deliveries outside Rome.

27. Landels, 214.

his appointment, in order to enhance his own image and, in contrast, to discredit the administration of his immediate predecessors under the Flavian emperors (101.2, 130.1).[28] In addition to general remarks about earlier laziness (*inertia ac segnitia*, 101.2; *neglegentia longi temporis*, 130.1), he even cites misappropriation of water revenues under Domitian (118.3).[29]

Such remarks must be viewed in perspective. Rome's aqueduct system did not collapse in the years preceding Nerva and Trajan, and there is ample archaeological evidence of Flavian upgrading and repair of the lines.[30] *De aquaeductu*, however, gives a strikingly different picture: the water system for which Frontinus has assumed responsibility is one much deteriorated from that of Augustan administration under Marcus Agrippa and his immediate successors, so much so that it demands the full attention of the new commissioner. Frontinus seeks to show his readers that he is not only knowledgeable about his new duties but is also taking charge, confidently making the rounds of the conduits, escorted not by lictors, as authorized by senatorial decree, but by his own *fides* and *auctoritas* (101.4). He specifies, in fact, how he has approached what he considers his greatest task—maintenance of the lines (17.2)—by having detailed maps made of the aqueducts (*formas ductuum*), thus gaining the advantage of being able to deal with a matter immediately, as if on the spot. He also describes the data depicted on these maps, including valleys crossed by the conduits, rivers, and sections on hillsides (17.3–4).

Frontinus' mapmaking merits closer attention. His decision to have plans of the system prepared should not surprise us, given his earlier military career and the Roman fondness for maps, which were frequently consulted by private individuals, used in schools, and directly employed in practical applications, both civil and military.[31] Indeed one of Frontinus' earlier works, which survives

---

28. *Aq*. 102.16–17 lists Manius Acilius Aviola as curator for the twenty-three years before Frontinus (A.D. 74–97), but the transmitted text may well be in error, reflecting the loss of several names during this period. Cf. Rodgers, "*Curatores Aquarum*," 177–80: "A series of ceremonial curators could with equal probability bear the blame for the deplorable legacy inherited by Frontinus" (177). Bruun, 178–79, suggests that Frontinus' list was never complete to begin with and that the office of *curator* may well have been vacant for some time before he assumed it.

29. Bruun, 209, describes the sum involved, 250,000 sesterces per year, as "ridiculously small."

30. Cf. Ashby, 72, 76, 104, 109, 119, 202–4, 209, 250, 285, 290, 291; Van Deman, 30, 38, 39–40, 48, 63, 64, 89, 203, 206, 229–30, 258–59, 310, 315, 316, 326; Bruun, 151–52.

31. For a comprehensive account of Roman mapmaking see O.A.W. Dilke, *Greek and Roman Maps* (Ithaca, N.Y., 1985). Hodge, 434 n. 5, observes that Frontinus' maps also included figures indicating sizes and distances.

only in fragments, was a treatise in two books on the art of surveying (*De arte mensoria*).[32] Frontinus' maps, however, appear to have done more than simply help him learn his job; they directly influenced the writing of *De aquaeductu*.

As already noted, his account of the development of the aqueduct system (4–16) is an unusual history, sketchy on background and dates but surprisingly complete in statistics concerning individual lines. Each chapter gives and locates the source or point of origin for the aqueduct, and the length of its course in Roman feet, both underground and above ground, with specific distances for sections on substructure or arcade.[33] For certain lines Frontinus also notes the point of confluence for subsidiary branches (5.6–8, 12, 15.4, 19.8–9), as well as the exact locations of settling tanks outside the city (19.1–2, 21.1–2).

We would not expect such details in a historical survey, but we can recognize their origin; Frontinus has copied his data from the maps he had prepared.[34] There are other indications that he has consulted his maps within *De aquaeductu*: he pinpoints the sources of individual lines through exact references to the existing road system outside the city (5.4, 6.5, 7.6, 8.2, 9.1, 10.5, 11.3, 14.1, 15.1). Frontinus' maps of the system therefore appear to have been quite detailed. Those sections describing individual lines in *De aquaeductu* are in many respects a *commentarius* to accompany the maps he cites.

What may have prompted Frontinus' approach to undertaking his duties? We do know that the *commentarii* of Marcus Agrippa were an important source and model for his published treatise, indeed no doubt Frontinus' inspiration for writing a *commentarius* of his own. Frontinus cites this work as the organizational basis of the *cura aquarum* established in 11 B.C. (98.1, 99.3).[35] Frontinus did more than simply follow the pattern of Rome's first water commissioner in

---

32. Bruun, 11, cites this expertise as a possible factor in Frontinus' appointment. The text of *De arte mensoria* may be found in C.O. Thulin, ed., *Corpus Agrimensorum Romanorum* (Stuttgart, 1913). In addition, Frontinus wrote other treatises on related topics (*De agrorum qualitate*, *De controversiis*, and *De limitibus*).

33. Frontinus' figures have been found to be generally consistent with the archaeological evidence. Blackman, "Length," 18, concludes that only the length he reports for the Anio Vetus is "seriously in error." Lanciani, 357, also comments on Frontinus' "mirabile precisione."

34. Cf. Grimal, xi–xii, who suggests that Frontinus has consulted earlier maps of the water system, perhaps prepared as early as Agrippa or when Claudius reorganized the system under imperial procurators. But if earlier maps existed, why would Frontinus make special mention of having prepared his own? Cf. Hodge, 172: "some of the 'plans' may have been diagrams of a known and existing network rather than true-to-scale and topographically accurate maps proper."

35. Meyer Reinhold, "The Imperial *Commentarii Aquarum*," *Classical Weekly* 28 (1935): 93. For the *commentarii* of Agrippa see J.-M. Roddaz, *Marcus Agrippa*, *BEFAR* 253 (Rome, 1984), 572–73; and Lanciani, 562.

preparing a written record of office. His decision to make his own maps of the system appears also to have been inspired by Agrippa himself: not by Agrippan maps of the water system proper, but by another, more famous achievement, the Agrippan map of the Roman world in the Porticus Vipsania.[36]

The most well-known example of monumental cartography in antiquity, Agrippa's map is cited by Pliny the Elder in recording the length of Baetica, a province in southern Spain, as he notes the discrepancies in distances reported by different authorities (*HN* 3.16–17):

> As for Agrippa, in a man of such great diligence and concern, especially in this undertaking when he proposed to display a map of the world for the citizens of Rome to look at, who would believe that he made a mistake, and, with him, the divine Augustus? For the emperor finished that portico containing the map begun by his sister, from the plan and notebooks [*commentarii*] of Marcus Agrippa.

This *locus classicus* on Agrippa's map provides important information about a project that Agrippa himself did not live to finish. He left behind memoranda, which became the source or authority on which the completed map was based. Within his *Natural History* Pliny quotes Agrippa some thirty times as a source for directions and boundaries, but he mentions the map itself only once more (*HN* 6.139). Pliny's thirty citations, whether based on the map itself or upon Agrippa's *commentarii,* indicate that the map was based on data gathered from the Roman road system; the statistics in the *Natural History* give lengths and widths of territory, distances from place to place, and boundary lines.[37] Agrippa's commentaries and the map completed from them were a compilation of data gained from his military experience, but once finished, the map had an educational function as well, displaying the world, of which Augustan Rome was the capital and focal point, to its citizens.[38] The map was also copied and circulated in abridged versions with commentaries, becoming an important source for later geographical treatises.[39]

---

36. The most recent discussion of Agrippa's map is that of C. Nicolet, *Space, Geography, and Politics in the Early Roman Empire* (Ann Arbor, 1991), 98–121. Cf. also Roddaz, 573–91; S.D.F. Detlefsen, *Ursprung, Einrichtung und Bedeutung der Erdkarte Agrippas* (Berlin, 1906); and A. Klotz, "Die geographischen commentarii des Agrippa und ihre Überreste," *Klio* 25 (1931): 35–58, 386–466. Dilke, *Greek and Roman Maps,* 41–53, gives a useful summary of the sources and problems.

37. Dilke, *Greek and Roman Maps,* 44–52, lists these citations, with an English translation.

38. Nicolet, 110–14.

39. Dilke, *Greek and Roman Maps,* 52–53.

Frontinus' maps were quite different in scale and purpose, but his decision to cite them in describing his preparation for office seems deliberate: he presents himself as modeling his administration on that of Rome's first water commissioner in his thorough and conscientious approach to his duties. Because Agrippa had prepared *commentarii* on the water system, Frontinus thought it appropriate to do the same. Presenting himself in *De aquaeductu* as a dedicated public servant determined to restore sound administration after years of neglect, Frontinus invites contemporaries to compare him to the celebrated Agrippa.

The most conspicuous comparison appears early in the work, in an apparently gratuitous digression on the Aqua Crabra near Tusculum (9.4–8): Agrippa had not tapped the Crabra for supply of the Aqua Julia because of local needs,[40] the watermen had illegally diverted its water for private gain, and Frontinus restored all of it to Tusculan consumers by order of the emperor. This action, concerning one potential source for one aqueduct, seems quite insignificant in itself, given the size of the overall system; moreover, supplies to the *ager Tusculanus* were not technically the concern of the *curator aquarum*.[41] Frontinus includes the incident specifically to draw attention to himself acting like Agrippa. The first water commissioner is cited seven more times within the treatise (9.9, 10.1, 25.1–2, 98.1, 99.3, 104.1, 116.3), but the episode of the Aqua Crabra seems deliberately chosen to color our first impressions of Frontinus as a senatorial commissioner and, by contrast, of the watermen whom he had inherited from earlier curators. His reference to them as *nostri aquarii* (9.6) seems scornfully ironic, in pointed contrast to the *singularis cura* that characterizes both Agrippa and himself in their approach to office (9.9).

Indeed there were parallels in the careers of Frontinus and his role model. Both men had demonstrated ability and potential through earlier military careers. Frontinus therefore presents in his treatise an image of himself as a loyal lieutenant of an enlightened *princeps* determined to correct the abuses of the past. Publishing his own *commentarius* as a treatise on the aqueducts was one way for Frontinus to demonstrate his commitment and competence. Preparing maps of the water system, on which the *commentarius* was partially based, was another.

As noted earlier, approximately one-third of *De aquaeductu* is devoted to

---

40. *Aq.* 9.5 cites two possible reasons why Agrippa did not tap the Crabra, but Frontinus deliberately expands upon the second, to emphasize Agrippa's *moderatio* and sensitivity to local needs (9.6), qualities that he himself notably demonstrates in this episode.

41. Bruun, 281–82 n. 58, points out that the Tusculan supplies were independent from the Roman lines. Hodge, 448 n. 17, observes that many of the Crabra's consumers were no doubt located at a distance downstream from the source.

citations of senatorial decrees, edicts, and laws of 11 B.C., shortly after the *cura aquarum* was established (99.1–4). Frontinus therefore presents himself as an administrator determined to restore the system to a more efficient management of the sort it enjoyed in earlier years, before Claudius' introduction of imperial procurators had made the office of water commissioner itself largely ceremonial (105).[42] Frontinus indeed says very little about the role of the procurators, mentioning them only twice in his treatise: in the case of private water grants, the procurator has the responsibility to insure that a *calix* is properly stamped (105.2), and Frontinus himself found evidence of illegal delivery for which the procurator was responsible (112.3).[43] In contrast, he spends considerable time listing all of his predecessors in office (102), not because his readers would be interested in the prosopographical details of earlier commissioners, but because his catalog demonstrates that they, like himself, were *principes civitatis*.[44]

Frontinus does more than present himself as a worthy successor to a distinguished list of public servants. Through a barrage of statistics on the aqueducts themselves, their courses above and below ground, pipe capacities, and delivery figures, he seeks to demonstrate his knowledge of details and aggressive approach to management (2.1–2, 119.3) of a system in which all citizens could take justifiable pride (16, 119.1). He asserts that his own efforts have already increased the capacity of the existing network through reclaiming water wasted or illegally procured (77, 87.2). Indeed, he proudly claims almost to have doubled the total delivery within the city (87.2–3).

Frontinus' figures on actual delivery deserve more scrutiny than they have received. Without new construction—and Frontinus in his short term as *curator* does not claim credit for any—so much extra water did not appear from nowhere. His actual achievements were important, but perhaps less dramatic than their presentation in *De aquaeductu* would suggest. As commissioner Frontinus had no doubt inherited imperial records that were inaccurate and out of date, perhaps never fully revised since the establishment of the *cura aquarum* in 11 B.C.[45]

Frontinus' own citations of volumes indicate varying states of completeness and accuracy in the records. For the first six aqueducts discussed he presents

---

42. Bruun, 153–89 and 207–17, presents the most recent full discussion of the *curatores* and imperial *procuratores*. Cf. also Ashby, 26–27; Grimal, xv, 94 n. 124; and Rodgers, "*Curatores Aquarum,*" 176.

43. Bruun, 207.

44. Bruun, 178.

45. Grimal, 86 n. 82, suggests that the figures in the record books cited by Frontinus reflect actual deliveries when the individual aqueducts were originally introduced, when it did not seem necessary to evaluate the potential capacity of each line.

almost formulaic accounts of capacity as given in the *commentarii* (65.1, 66.1, 67.1, 68.1, 69.1, 70.1). For the Aqua Alsietina he finds no records at all (71.1), and his citations for the Claudia and Anio Novus (72.1, 73.1) are given more tentatively, probably because the later Claudian lines were not as thoroughly documented in the record books.[46] The imperial records were therefore of doubtful value in determining the actual capacity of the system, and Frontinus set out to correct them. Some water was no doubt lost because of fraudulent deliveries, indeed probably a considerable amount,[47] but fraud alone cannot fully explain the huge discrepancies between Frontinus' own measurements, which he is careful to include (even though they are not listed in his table of contents), and the figures in the record books, sometimes less than half the capacity as he records it.[48]

In *De aquaeductu* Frontinus claims credit for the introduction of more accurate measurements, thereby making possible correction of the records and closer control over available supplies (74.4). He therefore states that his measurements made in the summer of A.D. 97 are accurate and reliable, in order to demonstrate both his diligence and administrative skill (69.2, 72.3, 74.3).[49] His claim to have compiled his treatise as a *formula administrationis* primarily for himself and successors must therefore be viewed as criticism, justified or not, of earlier commissioners and a challenge to all who would succeed him in office.

We must begin any study of water distribution in ancient Rome with *De aquaeductu*, but we must also recognize that it is a document presented to celebrate its author and the policies of the emperor who appointed him. Frontinus' presentation is selective, and our evidence suggests that the office of *curator aquarum* was quite unlike the picture he presents of a hardworking, "hands-on" administrator. Nor did later commissioners follow the model provided by Frontinus himself: there may well have been no (or very few) senatorial

---

46. Cf. Reinhold, "*Commentarii Aquarum*," 93. Rodgers, "*Copia Aquarum*," 359 n. 25, observes that there is no way to date or identify the figures in the records cited by Frontinus, and he seems skeptical of Frontinus' charge that earlier figures were unreliable (74.2).

47. Pliny *HN* 31.42 cites the *ambitio* and *avaritia* of suburban property owners in illegally diverting the Aqua Marcia and Aqua Virgo.

48. Ashby, 30–31, suggests that the official figures may well have reflected delivery at the terminal reservoirs in the city, not at the intakes of the aqueducts. Cf. Coarelli, "*Suburbium*," 43, who seems to take Frontinus' explanation of widespread fraudulent delivery at face value.

49. Robinson, "Water Supply," 70, cites Frontinus' "self-congratulation" on his measurements. For the date see Grimal, 86 n. 83: if Frontinus had taken his measurements in two consecutive summers, we might expect him to cite them accordingly, to strengthen his argument about capacity.

curatores aquarum throughout the second century A.D., until the principate of Septimius Severus, when ex-consuls appear in inscriptions as *curatores aquarum et Miniciae*, after the water commission had been merged with that for the grain supply.[50] *De aquaeductu* is therefore an anomaly, a treatise celebrating a position that in actuality never was what it describes.

We cannot approach *De aquaeductu* as a "simple and truthful narration of the facts,"[51] but we can exploit the evidence it gives us. Frontinus expects the statistics on water distribution that he cites to speak for themselves. Assessing how well they do is the task at hand.[52]

---

50. Bruun, 174–89. Cf. G. Rickman, *The Corn Supply of Ancient Rome* (Oxford, 1980), 253–56.

51. Bennett, xxv.

52. Bruun, 370, rightly observes that it will not do to take every passage and every word in *De aquaeductu* literally. It would also be a mistake to ignore what Frontinus does tell us. Cf. Robinson, "Water Supply," 78: "Although Frontinus' lists, as they have survived, are arithmetically inaccurate, it is clear that the water commissioners had in their office an adequate schematic account of the distribution for which they were responsible, as well as plans of the physical structures."

CHAPTER 3

The Aqua Appia

Although shorter and technically less sophisticated than later lines, Rome's first aqueduct was no doubt considered a remarkable achievement at the time of its introduction. Its construction required much more time than the normal eighteen-month term of Appius Claudius' censorship (5.3), and when completed the aqueduct earned its builder lasting fame (5.3; cf. *CIL* 11.1827; Cic. *Cael.* 14.34; Livy 9.29.5–6). Brought in from a low source, fifty feet below ground level at its intake (65.7),[1] the Appia entered Rome underground in the area of Spes Vetus (5.6), traversed the Caelian and Aventine (22.3), and ended at the bottom of the Clivus Publicii near the Porta Trigemina of the "Servian" Wall (5.9), near the northwest corner of the Aventine Hill in the southern Forum Boarium (fig. 1).

Because of its low level, the Appia's route within the city can be traced mainly from the evidence of Frontinus cited in the source note below. It is probable, however, that the course of the aqueduct under the Caelian, before it crossed to the Aventine, followed the ridge taken by later lines and paralleled by the *Arcus Caelimontani* of the Aqua Claudia. Antonio Colini attributed to it sections of solid tufa blocks enclosing a circular channel found in different sections of the Caelian.[2] Three sections of its *specus* have also been identified

---

Sources: *Aq.* 5; 18.7; 22.1, 3; 65; 79

1. This depth of the source is extraordinary, if indeed Frontinus' text is correct. Plautius' discovery of a source so far underground would seem to have justly merited the award of the name *Venox* (5.2). For the probable location see L. Quilici, *Collatia, Forma Italiae* I, x (Rome, 1974), 315–16. Frontinus' account of the source of the Aqua Virgo (10.3–4) gives an indication of the process involved: following the source of springs, which led to the supply itself. Cf. Hodge, 72–79.

2. A.M. Colini, *Storia e topografia del Celio nell' antichità, AttiPontAcc* ser. 3, vii (Rome, 1944), 82–83, 86–87; cf. A. Mucci, "Aqua Appia," in *Trionfo*, 30–32. The tufa

with remains of channels under the Little Aventine.[3] According to Frontinus, the Appia ran completely underground except for 60 *passus* on substructure and arcade in the immediate vicinity of the Porta Capena (*proximum Portam Capenam*, 5.5), in the valley between the Caelian and Aventine. This section is now lost, but Parker claims to have identified and photographed it in 1867, before it was destroyed.[4]

In comparison with later lines, the Appia was very basic in design, with no *piscina* installed along its conduit (22.1). Its course within Rome was also unusual, traversing much of the city underground and beginning its distribution only at a low point near the Tiber. The Appia therefore bypassed much of the city under which it ran, and, according to Frontinus, did not distribute at the Porta Capena where it emerged briefly from underground, an area where a *castellum* could easily have been installed.

We must recognize, of course, that from a purely technical standpoint, the Appia's low level prevented its distribution to higher areas (18.7); its elevation at its intake outside Rome has been estimated at no more than 24 meters above sea level, and its terminus at the foot of the Aventine at approximately 15 meters.[5] Its water therefore fell approximately 8 meters in a course of 11,190 *passus* (16.167 km), reflecting a drop over its course of not more than .5%, the minimum prescribed by Vitruvius in planning an aqueduct line (*De arch.* 8.6.1). Distribution at a higher point within Rome would have presented considerable problems, which Appius Claudius' engineers in the late fourth century B.C. might not have been able to surmount. However, distribution would certainly have been possible in the area of the Porta Capena, if with most scholars we accept Frontinus' account of the arcade and *substructio* there (5.5).[6] Indeed so little is known about the course of the Appia that there is little reason to reject Frontinus' notice. The section of 60 *passus* in the immediate vicinity of the Via Appia, Appius Claudius' highway and the other major construction of

---

blocks, or *lapides perterebrati* described by Vitruvius (*De arch.* 8.6.8), were interlocking, fixed to each other through stone collars. Other scholars have assigned the blocks to the *Rivus Herculaneus* of the Aqua Marcia; cf. Lanciani, 311; Ashby, 153–54; Van Deman, 134.

3. Ashby, 52–53; Van Deman, 26–27; Lanciani, 249–50, pl. I, 4a; Lugli, 2.336.

4. J.H. Parker, *The Aqueducts of Ancient Rome* (Oxford, 1876), 8–9, pl. XVIII, reproduced in Pace, 120, fig. 79. Frontinus' notice of a location *proximum Portam Capenam* indicates a line separate from the gate and "Servian" Wall, not incorporated into it. Cf. Lanciani, 249.

5. Ashby, 54; Grimal, 84–85 n. 74.

6. Lanciani, 249; Van Deman, 26; Grimal, 68 n. 9; Quilici, "Acquedotti," 62; Hodge, 412 n. 2. Cf. Ashby, 52–53, who argues that the Appia must have run underground at this point.

his censorship, would certainly have been the most conspicuous part of the entire line. The bridge itself, if dating from the original introduction of the line, would be one of the earliest, perhaps even the very earliest, use of an arcade in Roman architecture.[7] We might even expect a *castellum*, perhaps even of an ornamental nature, and distribution at that point as an architectural display intended to monumentalize Appius Claudius' censorship.

However, the nearby Fons Camenarum no doubt supplied the needs of the neighborhood in the late fourth century B.C. (cf. 4.2),[8] and there were other springs as well in the area, such as those that supplied the Piscina Publica.[9] It is the Appia's extension and terminus *ad Portam Trigeminam* that explain the reason for its introduction and the areas it originally served. Frontinus' notice of its ending *imo Clivo Publicio* indicates a low-level terminus near the northwest corner of the Aventine, certainly inside the Porta Trigemina, north of the present outlet of the Clivo di Rocca Savella, between the churches of S. Anna and S. Maria in Cosmedin.[10] From this point its water could have been distributed through pipes underground to the Forum Boarium, areas at the foot of the Aventine, the western Circus Maximus, and the southern part of the Campus Martius.[11] There is little reason to believe that there was ever a conduit on arches above ground at the terminal;[12] not only does Frontinus not mention an

---

7. Frontinus specifically mentions an arcade (*arcuatura*) at this location. A.W. Van Buren, "Wasserleitungen," *RE* 8A (1955): 468, suggests that this arcade was part of the second-century restoration of the Appia, replacing an earlier construction, perhaps of wood. Cf. Quilici, "Acquedotti," 69, who cites the second-century Aqua Marcia bridge from Capannelle to Spes Vetus as the first continuous arcade in Roman architecture. If part of the original aqueduct, Appius Claudius' arcuation might well have inspired the later, much more ambitious, bridge.

8. See Platner-Ashby, 89–90; Richardson, 63–64.

9. Lanciani, 234–35. For the Piscina Publica, a large swimming facility, see Platner-Ashby, 391–92; and Richardson, 292.

10. Lanciani, 251. If the aqueduct had terminated outside the gate—an arrangement highly unlikely, given its distribution inside—we would expect a notice of its location *extra portam*. The Porta Trigemina cannot be exactly located but must have stood near S. Maria in Cosmedin. For a discussion of the literary evidence see F. Coarelli, *Il Foro Boario dalle origini alla fine della repubblica* (Rome, 1988), 25–34; and Richardson, 310.

11. Hainzmann, *Untersuchungen*, 81.

12. So Lanciani, 312–13, followed by Ashby, 54; Grimal, 68 n. 9; and Hainzmann, *Untersuchungen*, 99 n. 3. Mucci, "Aqua Appia," 32, cites earlier evidence but reaches no conclusion. Lanciani connected the Arch of Lentulus and Crispinus, now destroyed (*CIL* 6.1385), with a branch line of the Marcia installed by Augustus; Ashby, 54, assigned it to an Augustan extension of the Appia. But the inscription on the arch as recorded indicates that it was a single span, not part of an arcade; it is more probably to be connected with a gate, possibly the Porta Trigemina itself, and therefore a parallel to the "Arch of Dolabella" on the Caelian. Cf. Colini, *Topografia*, 33; and Coarelli, *Foro*

arcade at the terminus, but such an arrangement, always rare within the walls of Rome, would have been quite impractical for an area as built up as the Forum Boarium.

Frontinus describes the immediate area of the terminus as the Salinae. The name itself is problematic: the term normally indicates saltworks or salt flats,[13] and such an installation may well have existed at one time in the early Forum Boarium, perhaps to indicate the spot where salt brought from the river mouth at Ostia was refined for use as table salt, or stored for shipment further inland along the Via Salaria or "Salt Road."[14] It is interesting to speculate whether such processing went on in the southern Forum Boarium as late as the time of the Appia's introduction; certainly the refining and manufacture of salt would have required a supply of fresh water.[15] This is unlikely, however; the manufacture of salt in the vicinity had probably long been ended by the late fourth century, given the development of the immediate area and its partial enclosure by the circuit of the "Servian" Wall, but the name Salinae seems to have remained.

Frontinus' identification of the spot (*qui locus est ad Portam Trigeminam,* 5.5; *ad Portam Trigeminam, qui locus Salinae adpellatur,* 5.9) suggests that the Salinae were little more than a topographical memory at the time he was writing. Livy mentions them in his account of a fire in the Forum Boarium in 213 B.C. that destroyed everything between the Salinae and the Porta Carmentalis (24.47.15), evidence that the Forum area was heavily built-up at that time, but even in this context they appear to be little more than a topographical point of reference.

The Salinae were only one installation near the Appia's terminus. The Republican port of Rome was to be found in the Forum Boarium, immediately south of the Tiber Island and adjacent to the Pons Sublicius, the first bridge across the Tiber and probable site of the earlier ferry across the river. This area

---

Boario, 45–50, who argues for identification of the arch with the Porta Trigemina. C. Buzzetti, "Arcus in platea Sanctae Mariae," *Bolletino della Unione Storia ed Arte* 77 (1984): 24, suggests that it is "not improbable" that the arch, reconstructed in the Augustan period as a monumental gate, also carried the branch line of the Marcia that supplied the Transtiber to the Pons Aemilius.

13. Cf. Jordan, 1:235; H. Blümner, "Salz," *RE* 1A2 (1920): 2078–79.

14. Platner-Ashby, 462, describe the Salinae as warehouses for salt brought from Ostia. Richardson, 341, points out that there are no references to them as *horrea* in our literary sources, something we would expect if they were storage buildings. Coarelli, *Foro Boario,* 110–11, suggests that the name Salinae marks the terminus of the Via Salaria within the Forum Boarium; from here salt brought from the river mouth would have been shipped inland.

15. Cf. Pliny *HN* 31.81; and Forbes, 3:164–209.

had taken on an ever increasing commercial importance as early as the fourth century B.C. The number of cults of divinities with commercial character, such as Portunus, Fortuna, Mater Matuta, Hercules, Diana, and Ceres, Liber and Libera, points to its role as a port and marketplace.[16] It therefore seems likely that Appius Claudius' aqueduct was brought specifically to the district to meet pressing needs that could not be supplied by existing wells or cisterns.

The northern Forum Boarium was probably adequately supplied from natural sources. The spring of the Lupercal stood not far from the base of the Palatine in the Velabrum, and the Forum Romanum had both the Fons Juturnae and the Tullianum spring.[17] The southern Forum Boarium, however, appears to have lacked a water source nearby. The only recorded spring in the immediate vicinity, the Fons Scaurianus, located by inscriptional evidence near the church of S. Prisca on the Aventine,[18] was too distant to furnish the water needed close to the river. One particularly good parallel of water being brought to a Roman port facility is the aqueduct at the port of Cosa, a Latin colony established in the early third century B.C.[19] Indeed the arrangement at Cosa may well have been inspired by Appius Claudius' own planning in the capital.

The location *imo Clivo Publicii* has still another significance. According to our literary evidence, the street was constructed and paved only in 240 B.C.,[20] but it probably followed what was already a well-established natural route from the area along the river to the higher ground on the Aventine.[21] The Appia's water was therefore distributed along a thoroughfare leading down from the hill either within or through the Porta Trigemina into the Forum Boarium. From its terminus its water met public needs in a growing commercial area. Appius Claudius' introduction of the aqueduct therefore complemented the other major achievement of his censorship, the building of the great south road linking Latium and Campania, which served commercial as well as military aims. Although our evidence on Appius Claudius' censorship has recently sparked

---

16. Coarelli, *Foro Boario*, 107–56, gives a summary of its early history. Cf. also A.M. Colini, "Il porto fluviale del Foro Boario a Roma," in J.H. D'Arms and E.C. Kopff, eds., *The Seaborne Commerce of Ancient Rome: Studies in Archaeology and History, MAAR* 36 (1980): 43–51.

17. Lanciani, 225–26, 233, 235–36.

18. Platner-Ashby, 211; Richardson, 153.

19. Cf. Quilici, "Acquedotti," 64; and A.M. McCann, *The Roman Port and Fishery of Cosa: A Center of Ancient Trade* (Princeton, 1987), 92–93, 98–128, 328–29, 342.

20. Platner-Ashby, 124; Richardson, 90; Coarelli, *Foro Boario*, 29–34.

21. *LFUR*, pl. 34, identifies the Clivus with the present Clivo di Rocca Savella, which terminates near the Tiber at the Church of S. Anna. Coarelli, *Foro Boario*, 29–34, argues that it is rather to be associated with the Clivo dei Publicii, which ends at the Via della Greca.

scholarly debate, his political program points to an interest in implementing social and economic changes necessary to permit Rome to assume a more important position in Italian commerce.[22] Construction of the Aqua Appia was certainly consistent with such goals; indeed it appears to have played a large part in achieving them.

Our knowledge of the Appia's distribution is based entirely on Frontinus (79.1–2), evidence that reflects the aqueduct's role within the city's water system only at the end of the first century A.D. Reconstruction of its original distribution is therefore conjectural, but worth exploring further. The southern Forum Boarium in the late fourth century clearly needed additional water; in no other way can we explain the extraordinary measures taken by Appius Claudius' engineers to tap a source at such a level and bring the Appia to the district from such a distance.

One other potential need comes to mind: supply for the nearby Circus Maximus. Frontinus himself cites restrictions governing delivery to the Circus (97.1–2). The Circus, of course, had its own natural water supply through the Marrana brook,[23] and its water needs do not appear to have been acute in the late fourth century, when it was still a generally open market area reserved for racing on particular days. It is worth noting that its first starting gates (*carceres*), probably of wooden construction, were installed only in 329 B.C., shortly before the Appia itself was introduced (cf. Livy 8.20.2). Unlike the monumental Circus of the Augustan period and empire, it would have required little or no aqueduct water at this time.[24] The commercial needs of the Forum Boarium itself were therefore the key reason for the introduction of the Appia.

From Frontinus we get a better idea of how water from the Appia was distributed almost four hundred years later, a period when Rome had been radically transformed, the Appia itself had been reworked three times (7.3, 5.6–8), and the city's water needs had grown enormously. Frontinus states that the

---

22. Cf. E.S. Staveley, "The Political Aims of Appius Claudius Caecus," *Historia* 8 (1959): 410–33; E. Ferenczy, "The Censorship of Appius Claudius Caecus," *Acta Antiqua* 15 (1967): 27–61; and B. MacBain, "Appius Claudius Caecus and the Via Appia," *CQ* 30 (1980): 356–72. Coarelli, *Foro Boario*, 111 n. 6, suggests that the nearby Ara Maxima was also a factor in Appius Claudius' decision to introduce the aqueduct. During his censorship the cult of Hercules at the Ara Maxima was transferred from private to public control (Macr. *Sat*. 3.6.13), an action not inconsistent with efforts to develop the nearby area of the port.

23. Platner-Ashby, 114; Richardson, 84.

24. Cf. J.H. Humphrey, *Roman Circuses: Arenas for Chariot Racing* (Berkeley and Los Angeles, 1986), 67–72. A notice in Livy (7.3.1–2) indicates that the Circus was subject to flooding as late as 363 B.C., further evidence that it was still largely an open area, with no permanent installations.

Appia supplied seven of the fourteen Augustan regions: the Caelian (II), Roman Forum (VIII), Circus Flaminius (IX), Circus Maximus (XI), Piscina Publica (XII), Aventine (XIII), and Transtiber (XIV).

The Appia's distribution to Regions II, XII, and XIV, all some distance from its terminus *ad Portam Trigeminam*, is surprising, given the low level of the line. The notice in Frontinus indicates, however, that at one or more later periods secondary branch lines must have been laid to low-lying parts of the Caelian, Piscina Publica, and the Transtiber, the last probably by a conduit over the Pons Aemilius. We might associate these extensions with revisions of the aqueduct system by Q. Marcius Rex in the late first century B.C. (7.3), or with much more probability, with the revisions of the late first century B.C., either the work of Agrippa in 33 B.C. (9.9) or that of Augustus, who increased the Appia's capacity by giving it a supplemental source, the *ramus* bearing the name *Appia Augusta* (5.6–8).

The length of the Augustan branch, more than half that of the original conduit of the Appia (6,380 *passus* vs. 11,190), indicates that it was an extremely ambitious addition to the aqueduct. Frontinus pinpoints the location of its source in great detail (5.7),[25] but its course within the city is unknown, as is its juncture with the main conduit at a location Frontinus calls *ad Gemellos* (5.6).[26] Attempts have been made to identify the *ramus Augustae* with the remains of a channel running underground at Spes Vetus,[27] but the evidence is inconclusive, complicated in particular by problems in levels. There is no evidence that Augustus' line constituted a separate branch of the Appia within the city supplying the Esquiline or other regions.[28] Not only does Frontinus omit the Esquiline from the list of regions to which the Appia distributed, but Augustus' construction seems rather to have been planned to augment the Appia's capacity at its terminus through the delivery system already in place. In this way the additional supply could serve increased needs in the immediate area and make possible extension of the aqueduct through secondary lines to places more distant from the Forum Boarium.

Yet even with the addition of its Augustan supplement, by Frontinus' time the Appia played only a secondary role within the city's overall water system. Its 699 *quinariae* constitute only 7% of the total volume of water delivered in

---

25. For the probable source see Quilici, *Collatia*, 257–58.

26. The term "Twins" seems to have been a nickname used by the watermen. Cf. Platner-Ashby, 246; Richardson, 180–81; Pace, 239 n. 4; and Rodgers, "Textual Temptations," 131.

27. E. Gatti, "Avanzi di acquedotti romani scoperti presso Porta Maggiore," *BullCom* 40 (1922): 232–33; cf. Ashby, 82.

28. So Lugli, 2.338, followed by Panimolle, 46.

Rome (9,955 *quinariae,* 78.3). Moreover, its 20 *castella* reflect only 8% of a total of 247 within the city. Analysis of its distribution, as reported by Frontinus, indicates that Appia mainly supplied public basins and buildings:

Appia distribution *intra urbem*	1) *nomine Caesaris*	151 q.	(21.6%)
	2) *privatis*	194 q.	(27.5%)
	3) *usibus publicis*	354 q.	(50.8%):
	a) 1 *castra*	4 q.	(.6%)
	b) 14 *opera publica*	123 q.	(17.6%)
	c) 1 *munus*	2 q.	(.3%)
	d) 92 *lacus*	226 q.	(32.3%)
	TOTAL	699 q.	

Little more than a fourth of the Appia's water was distributed to private citizens. This figure is not surprising, given the early date of its introduction and its low level. Frontinus states that water in the early Republic was brought in solely for public use, with the exception of grants to a few prominent citizens made with general consent (94.3, 94.5–6). Moreover, while elegant houses could be found in all neighborhoods of ancient Rome, we would expect most *privati* who enjoyed the amenity of their own water supply, both in the early Republic and at Frontinus' time, to have resided at higher elevations that the Appia could not reach.

Instead the figures show that the Appia delivered over 70% of its water to imperial buildings and for public uses. Yet the small number of *quinariae* involved (only 8.7% of the total distributed by all the aqueducts for such purposes) indicates that the Appia must have supplied only small buildings or installations. Despite its proximity to the Circus Maximus, it seems to have furnished very little water, if any, to the complex, even after the Circus had undergone major reconstruction in the late Republic.[29] The fourteen public works to which it supplied 123 *quinariae* could not have been major monuments, perhaps small baths (*balneae*) or installations in the vicinity of the river.

One observation is worth making: the Appia furnished almost a third of its total capacity, 226 *quinariae,* to 92 *lacus,* a delivery to public basins greater in actual volume than that of the Anio Vetus (218 *quinariae* to 94 *lacus,* 80.2) and in fact almost as great as that of the Aqua Marcia (256 *quinariae* to 113 *lacus,* 81.2), both later lines with volumes much greater than that of the Appia itself. This again is not surprising in light of Frontinus' notice that earlier lines

29. For Augustan monumentalization of the Circus see Humphrey, 73–100.

were introduced to meet public needs. But Frontinus' data are valuable here for another reason: they indicate an average delivery of 2.45 *quinariae* to each *lacus,* a figure fairly standard for all the aqueducts Frontinus records,[30] and permit us to evaluate other statistics concerning the Appia's distribution, specifically the volume of water delivered to the one camp and one *munus* listed.

Archaeological evidence from Pompeii and Herculaneum indicates that public *lacus* were small structures of fairly standard capacity, usually located at intersections to serve particular neighborhoods.[31] If the basins supplied by the Appia received an average of 2.45 *quinariae* each, the one *castra* to which it furnished only 4 *quinariae* must have been quite small; the most likely installation might well have been the headquarters of the VIIth cohort of the *vigiles,* which appears to have had responsibility for Regions XI and XIV.[32] Even more telling is the figure of 2 *quinariae* supplied to one *munus,* which must have been a single imperial donation, perhaps to be connected with the Circus Maximus: if this were an ornamental fountain, it must have been quite modest. This too may reflect the Appia's low level and early date. Decorative fountains do not seem to have been introduced into Rome until the first century B.C., and the Appia's level was probably too low to supply any large display of water near its terminus.

Frontinus' evidence therefore suggests that by the late first century A.D. the Appia served largely utilitarian purposes, supplying a limited volume of water to low-lying areas of the city, perhaps functioning as a backup line in many of the regions to which it was extended, to furnish a substitute supply if other conduits were interrupted (87.5). Frontinus does specifically mention his measurement of the Appia *ad Gemellos* near Spes Vetus (65.2–4), citing large discrepancies between the volume actually found (1,825 *quinariae*) and the figures in the record books (841 *quinariae*), as well as the delivery figures cited in his detailed analysis of distribution (699 *quinariae,* 79.2). These discrepancies are considerable, indicating a capacity for the Appia twice that listed in the official records and two and half times its scheduled delivery.[33] Yet Frontinus' explanation

30. Lanciani, 579; Bruun, 102.

31. Cf. Richardson, *Pompeii,* 57–59; Hodge, 304–7; B. Gockel, "Laufbrunnen im römischen Pompeji," in *Wasserversorgung,* 194–95.

32. See P.K. Baillie-Reynolds, *The Vigiles of Imperial Rome* (Oxford, 1926), 55–58; one of its *excubitoria,* dating from the time of Hadrian, has been excavated in Trastevere. Cf. Rainbird, 154–56.

33. We might speculate whether the *commentarii* cited by Frontinus (65.1) reported delivery figures for the Appia *before* introduction of the *ramus Augustae* dramatically increased its capacity. Yet we would expect the records to have been brought up to date following the Augustan improvements, especially if such large figures were involved. Cf. Reinhold, "*Commentarii Aquarum,*" 93.

of them seems unconvincing, given the differences in the figures themselves and the limited distribution of the Appia: he attributes the difference to leaks in many sections of the city (*in pluribus urbis partibus,* 65.5), as well as to certain illegal taps (*quasdam fistulas intra urbem illicitas,* 65.6).[34] Frontinus does not appear to be particularly interested in the Appia. Because of its low level and therefore somewhat limited role in the overall system, he might well have considered it impractical to repair the leaks he found.

Belisarius is recorded to have cut Rome's aqueducts in the sixth century A.D., to prevent their use by enemies as a means of infiltrating the city (Proc. *Goth.* 5.19.8). Yet the Appia, being the lowest line on the eastern side of the Tiber and therefore the most difficult to block, may have continued to function well into the Middle Ages. The eighth-century Einsiedeln Itinerary (13.22) lists an *aqua subtus montem Aventinum currens* near the *ecclesia Graecorum* (S. Maria in Cosmedin), which would certainly have been in the area of its terminus.[35]

---

34. Mucci, "Aqua Appia," 31, attributes this leakage to the sections of tufa conduit found on the Caelian and assigned by Colini to the Appia (cf. n. 2, above).

35. Ashby, 50. The Itinerary, however, might instead refer to the Aqua Marrana. Cf. A. Piganiol, "Les origines du Forum Boarium," *MEFRA* 29 (1909): 129–30; and A. von Gerkan, "Der Lauf der römischen Stadtmauer vom Kapitol zum Aventin," *RM* 46 (1931): 168–69.

CHAPTER 4

The Aqua Anio Vetus

Although it was introduced only forty years after the Aqua Appia, the Anio Vetus was a much more ambitious undertaking. It drew its water from a much higher elevation in the upper Anio Valley and ran a course almost four times the length of the Appia. Despite the impossibility of determining the actual volume of its delivery, the Anio Vetus certainly had a total capacity far greater than that of the earlier aqueduct. The time required for its construction must have been much longer than that of the Appia: Frontinus states that the line was far from finished two years after the original contracts had been let (6.2). The costs also have been considerably greater, since he reports that spoils from the war with Pyrrhus furnished the necessary revenues (6.2).

In some respects the Anio Vetus was similar to the Appia in overall design and introduction: it too ran almost entirely underground, except for 221 *passus* on substructures, in part to maintain a constant gradient in its fall throughout its meandering course (6.6), but also because of military considerations, to prevent enemies from intercepting an exposed conduit outside Rome (18.4–5). Like the Appia, it also approached the city underground in the vicinity of Spes Vetus and distributed its water just inside a gate of the "Servian" Wall (21.3). But in other respects the Anio Vetus was technically a much more sophisticated line. By the time of Frontinus it had a *piscina* incorporated into its conduit (21.1), drew some of its water from another aqueduct, the Marcia (67.3), and supplied a branch line of its own, the *specus Octavianus* (21.2).

In contrast to the Appia, the line of the Anio Vetus can be traced within the city through physical remains of its channel and *cippi* erected during an Augustan restoration. Its *specus* was discovered below ancient ground level just

---

Sources: *Aq.* 6; 13.5; 18.6; 21; 66; 67.3; 80; 92–92

inside the Porta Maggiore;[1] from here it continued to run north-northwest along the line of maximum elevation on the Esquiline, turned west-southwest near the Porta Viminalis, under the present Stazione Termini, and then continued in a southeast direction to the Porta Esquilina, just west of Piazza Vittorio Emanuele, a roundabout course probably chosen to avoid the lower ground of the Villa Altieri situated between the Porta Maggiore and Porta Esquilina (fig. 1).[2] Here, according to Frontinus, it began its distribution in deep channels (*in altos rivos*, 21.3).[3] Recent excavations under the church of S. Vito have brought to light the *specus* of the aqueduct and remains of a *castellum* near the Porta Esquilina.[4]

The aqueduct's terminus inside the Porta Esquilina, an entry point considerably above the level of the Appia at the Porta Trigemina, and its much greater capacity made possible its delivery through large areas of Rome. Frontinus attributes to it 35 *castella*, a number almost twice that of the Appia, which indicates more widespread distribution. Indeed because of the elevation of its source in the upper Anio valley, the Anio Vetus could have delivered water to the highest points of the city, a fact confirmed elsewhere by Frontinus (18.6). But as Frontinus indicates in this same passage, although the Anio Vetus could have been brought over valleys and low ground between the hills, it was not so carried and therefore did not supply the *loca altiora* of Rome. Confirmation of its lower level is also to be found in another notice of Frontinus that discusses the muddying of aqueduct water after storms: because the Anio Vetus ran at a level lower than those of other lines, it kept its pollution to itself (91.1). The evidence therefore suggests that it was distributed at a fairly low level to those areas that it could reach from the Porta Esquilina by gravity flow.

Frontinus' catalog of the areas it served (80.2) must therefore be analyzed in light of its terminus and distribution. He lists the Porta Capena (I), Isis and Serapis (III), Templum Pacis (IV), Esquiliae (V), Alta Semita (VI), Via Lata

---

1. Gatti, 228–32; Van Deman, 54–55, pl. XII.

2. For the course see C. Roncaioli Lamberti, "L'Aqua Anio Vetus," in *Trionfo*, 33–40; Ashby, 81–86; Van Deman, 54–57; Hainzmann, *Untersuchungen*, 87; Lanciani, 262–64.

3. *Altos* can be translated "deep" (Ashby, 55; Grimal, 16) or "high" (Bennett, 363; "high lying," Van Deman, 57; Pace, 254). Grimal, 80 n. 61, observes that the distribution of the Anio Vetus in the city was underground, which gives greater support to "deep" channels, especially when we compare later aqueducts, particularly the high-level Anio Novus. The term *rivos* may be significant as well, perhaps an indication of distribution through free-flow channels, rather than a pipe system. On the extent of such channels within the city see Bruun, 118–23, who argues that they were probably more widespread than generally supposed.

4. V. Santa Maria Scrinari, "Brevi note sugli scavi sotto la chiesa di S. Vito," *Archeologia Laziale* 2, *QArchEtr* 3 (1979): 61–62.

(VII), Forum Romanum (VIII), Circus Flaminius (IX), Piscina Publica (XII), and Transtiber (XIV) as regions supplied by the Anio Vetus in his day. But for which of these ten regions was the aqueduct originally intended? It seems quite unlikely that its water would have been carried across the Tiber before a much later period, or even that Region VII in the Campus Martius or Region VI on the Quirinal would have required an outside water supply as early as the time of its introduction. In the third century the Campus Martius was still largely an open area, and both districts would certainly have had access to the Fons Cati or other springs on the Quirinal.[5] Distribution of the Anio Vetus to these regions certainly reflects extensions through branch lines at later dates.

Frontinus indicates that the Forum Romanum (VIII) and Circus Flaminius (IX), two regions in the city center, were served by both the Anio Vetus and Appia (79–80).[6] If water from the Anio Vetus was brought to them shortly after the line's introduction into Rome, we might interpret this distribution as evidence of increased needs in the center, particularly in the Subura, which the Appia alone could not satisfy or could not reach, given its low level and terminus near the Tiber. What is most striking, however, about the delivery of the Anio Vetus is the number of eastern regions Frontinus cites: the Porta Capena (I), Isis and Serapis (III), Templum Pacis (IV), and Esquiliae (V) were all supplied by the Anio Vetus. Distribution to these areas would not have presented serious technical problems; all of them, at least in part, lie at a ground level below that of the Porta Esquilina.

Whether all of them were included in the original distribution plan cannot be determined and may in fact be unlikely, given the line of the "Servian" Wall, which would have discouraged building in areas outside its perimeter, at least at the time of the aqueduct's introduction and probably well into the Republic. But all these regions represent eastern areas of the city previously not supplied by an aqueduct. The Esquiline Hill itself, although not totally devoid of natural sources of water,[7] did not have a major spring associated with it; increased inhabitation in the area no doubt necessitated introduction of a new water supply.[8] The point of entry for the Anio Vetus and its distribution, as presented by Frontinus, are therefore good evidence that by the early third century B.C. the city had begun to expand eastward to higher areas.

---

5. Platner-Ashby, 109; Richardson, 79; cf. Lanciani, 236, citing other springs on the Quirinal from which the Porta Fontinalis of the "Servian" Wall got its name.

6. Livy *Per.* 54 cites a dedication *in aede* of an "Aqua Anio," presumably the Anio Vetus, a notice too vague to be of any significant topographical value.

7. See Lanciani, 237–38.

8. Cf. P.A. Brunt, *Italian Manpower, 215 B.C.–A.D. 14* (Oxford, 1971), 384.

One other observation should be made about its distribution. According to Frontinus, the Anio Vetus did not distribute to the Caelian (II), Palatine (X), Circus Maximus (XI), or Aventine (XIII). In light of Frontinus' notice about levels (18.6), we would not expect delivery to the Palatine, certainly one of the *loca altiora* within Rome, or to the Caelian, which was in large part separated from the terminus of the Anio Vetus by an intervening valley. The Circus Maximus and Aventine were even more distant from the Porta Esquilina, and were also among the regions already served by the Appia. In this way the distribution of Rome's second aqueduct appears to have been carefully planned to complement that of the first, as well as to meet existing needs.

The Anio Vetus underwent three known restorations at later periods, the first two carried out by Q. Marcius Rex in 144 B.C. and Agrippa in 33 B.C. (7.3, 9.7), and a third by Augustus in 11–4 B.C., attested by the only extant *cippi* associated with the line.[9] Augustus is also generally associated with construction of the *specus Octavianus*, a branch line described by Frontinus (21.2).[10] There is little evidence for an Augustan date for this branch except its name,[11] which is itself problematic: if we are to associate construction of the branch with the emperor's restoration of the Anio Vetus in 11–4 B.C., we might expect either a notice to this effect from Frontinus, like that concerning the *ramus Augustae* of the Appia (5.6), or perhaps another name for the *specus* that would reflect the emperor's official title after 27 B.C.[12] Instead Frontinus reports only that the branch was called *Octavianus (in specum qui vocatur Octavianum*, 21.2), a term that seems informal, perhaps little more than a nickname.

The course and purpose of the *specus Octavianus* are equally problematic, since Frontinus' topographical notices cannot be confirmed through known archaeological remains. Its point of divergence from the main conduit is unknown, as are the Horti Asiniani in the district of the Via Nova where it was distributed. Traces of a channel of Augustan reticulate have been found running to the line of the Aurelian Wall between the Amphitheatrum Castrense and the Porta Latina; from this evidence Lanciani suggested that the branch ran southwest from the main line of the Anio Vetus along the Aurelian Wall

9. Ashby, 55, 57; Van Deman, 29–30; Lanciani, 258.

10. Ashby, 55; Hainzmann, *Untersuchungen*, 86; Van Deman, 66; Lanciani, 264; Robinson, "Water Supply," 59. Roncaioli Lamberti, 40, tentatively assigns the branch to Agrippa.

11. The *specus* might well be associated rather with an Octavius of the late Republic, although no Octavii are attested as censors in this period. Certainly the length of the line would point to considerable expense for its construction.

12. Cf. the name *aqua Augusta*, given to a supplemental branch of the Marcia added by the emperor (12.1) and also to the Aqua Alsietina (11.1).

toward the Porta Latina, where it turned northwest to run along the Via Appia in the direction of the Porta Capena, perhaps crossing the Via Appia itself on an arch to be identified with the Arcus Drusi (not to be confused with the present "Arco di Druso" near the Porta S. Sebastiano of the Aurelian Wall), and then was distributed around the area later occupied by the Baths of Caracalla.[13]

This suggestion, although generally accepted by later scholars, is not without difficulties. To connect Frontinus' notice of the Via Nova with the area near the Porta Capena we must conclude that the topographical reference is a later interpolation into the text;[14] the Via Nova itself was not built until some one hundred years after Frontinus, in the early third century A.D.[15] Since the Horti Asiniani are otherwise unattested, the passage may refer to quite another area of the city. However, there may well have been good reason to run a branch line southwest from the main channel of the Anio Vetus to the southeast area of the city, near the Little Aventine.

Frontinus reports that Regions I and XII were supplied by the Anio Vetus (80.2), no doubt through the *specus Octavianus.* Because the terminus of the main conduit was at the Porta Esquilina, some distance to the northeast, with the Caelian Hill immediately to its south, the Anio Vetus could be brought most easily to the areas of the Porta Capena and Piscina Publica through a branch deviating from the main conduit outside the city. Assigning the *specus Octavianus* to the Agrippan or Augustan overhaul of the water system, to serve growing areas outside the circuit of the "Servian" Wall, therefore seems quite reasonable. The supplement of 164 *quinariae* given by the Marcia to the Anio Vetus (67.3) might also be dated to this period, specifically for supply of this branch line.[16]

Still another problem is that of the capacity and distribution of the Anio Vetus in Frontinus' time, a question complicated by the conflicting figures given by the manuscripts. In his discussion of his measurements Frontinus states that the Anio Vetus distributed a total of 1,608 *quinariae,* both above and below its *piscina* (66.6). Moreover, in this same chapter (66.2) he cites a discrepancy of 2,857 *quinariae* between his own measurements at the intake and the figures in the record books (more than 50% of the actual capacity, according to his

13. Lanciani, 264–67.

14. Grimal, 79 n. 60, describes this reference as a "retouche topographique" (cf. Grimal, 70 n. 14); Hainzmann, *Untersuchungen,* 86 n. 3.

15. Cf. Platner-Ashby, 565; Richardson, 417.

16. Grimal, 85 n. 75, suggests that the Marcia's supplement was immediately distributed outside the city, perhaps to the Tiburtines mentioned in 66.2.

measurements), and a loss of 2,788 *quinariae* along the entire line, which he attributes not to inaccurate measurements but to illegal diversion (66.7). Yet the totals he gives elsewhere in his discussion of its delivery (80) are again different, reflecting a distribution of 2,081½ *quinariae*, 573 *extra urbem* and 1,508½ *intra urbem*.

In addition, Frontinus' statistics on the distribution within Rome, which follow immediately in this same chapter, give still another figure. Without emendation of the text to make sense of the arithmetic, it seems best to base our analysis on the total obtained from the most detailed discussion Frontinus provides, his breakdown of distribution that appears in 80.2:

Anio Vetus distribution	1) *nomine Caesaris*	64½ q.	(5.8%)
intra urbem	2) *privatis*	490 q.	(44.3%)
	3) *usibus publicis*	552 q.	(49.8%):
	a) 1 *castra*	50 q.	(4.5%)
	b) 19 *opera publica*	196 q.	(17.7%)
	c) 9 *munera*	88 q.	(7.9%)
	d) 94 *lacus*	218 q.	(19.7%)
	TOTAL	1,106½ q.	

There is another reason for following the total given in the above analysis. The delivery of 1,106½ *quinariae intra urbem* that it records, if added to the 573 distributed *extra urbem* (80.1), results in a total of 1,679½ *quinariae*, a figure only 69½ greater than that which Frontinus gives earlier in his notice on measurements, 1,348 *quinariae* below the *piscina* (66.6), plus 262 *quinariae* above it (66.3). While this near-agreement cannot inspire complete confidence in the accuracy of the statistics as reported, it results in a margin of error of approximately 4%, which, given the limited evidence available, appears acceptable for analysis of the delivery.

In examining the above statistics, we are immediately struck by differences from the delivery of the Appia: very little of the Anio Vetus seems to have supplied imperial buildings, only 5.8% of its total distribution. However, almost half of its water was distributed to *privati*, a figure reflecting its higher entry into the city, which made possible delivery to residential areas on the eastern hills.

Approximately 50% of its total capacity went to public works, about the same ratio as that of the Appia. Frontinus' statistics, however, point to important differences, not only in the larger volumes involved but also in his detailed analysis of their distribution. Like the Appia, the Anio Vetus supplied one camp, but the much greater volume distributed to this installation, 50 *quinariae* instead of 4, indicates that it was a complex much larger in size; no single installation can be identified as the most likely to have been supplied. It also supplied 196 *quinariae* to 19 public works, an average of 10 *quinariae* to each, about the same as that of the Appia. We should note as well its 88 *quinariae* to 9 *munera*, an average of 10 *quinariae* to each. The higher elevation of the line therefore permitted a much greater supply to fountains than the single *munus* in the Appia's distribution.[17]

Even more significant is its limited distribution to *lacus:* the Anio Vetus furnished only 218 *quinariae*—less than 20% of its total capacity—to public basins, 94 in all. Frontinus' data give an average delivery of 2.32 *quinariae* per *lacus,* a figure quite close to that seen in the Appia's statistics, another reason for placing confidence in the numbers given in his breakdown of the distribution in chapter 80.[18] But in comparison with that of the Appia, the distribution of the Anio Vetus to public basins was much smaller, no doubt because of the quality of its water. Anio water could become muddy after storms, a problem Frontinus discusses at length (90–92). As a result, he states that the Anio Vetus was reserved whenever possible for industrial uses and irrigation (92), thereby freeing the Marcia and better water for drinking. Many of the basins it supplied may well have been troughs for watering animals rather than public taps for drinking water.[19]

This infiltration of mud is an unusual feature in the Anio Vetus: we might ask why the water commissioners persisted in keeping the line active, despite this defect, or why they took no steps to improve the quality of its water by tapping other springs in the upper Anio valley to which the aqueduct could have been extended. Frontinus cites planning by Nerva for such an improvement of the Anio Novus (93), but none for the Anio Vetus. Any such reworking might have adversely affected the supply delivered for Tibur (6.5, 66.2), although that too would not have been an insurmountable difficulty, given the abundance of other sources in the upper Anio valley. The Anio Vetus was an older line, and was probably not considered worth a major overhaul.[20] After the intro-

---

17. Pace, 122, suggests that the Anio Vetus had a terminal fountain at the Lacus Orphei, near the Porticus Liviae. Cf. Platner-Ashby, 313–14; Richardson, 231; and E. Rodriguez Almeida, "I confini interni della 'Regio V,' Esquiliae, nella 'Forma Urbis Marmorea'," in G. Pisani Sartorio and L. Quilici, eds., *Roma Capitale 1870–1911. L'archeologia in Roma capitale fra sterro e scavo,* Exhibition Catalog, Venice, 1983, 171–72, who dates the Lacus Orphei to the Augustan period, and notes that it is probably to be connected with Augustus' reworking of the Anio Vetus.

18. Lanciani, 579; Bruun, 102.

19. For a good example of such a watering trough see R. Bartoccini, "Il rifornimento idrico della Colonia *Julia Felix Lucus Feroniae,*" *Autostrade* 7–8 (1963): 9–10, fig. 24.

20. Recent excavations have confirmed, however, that its distribution system in the

duction of the Claudian aqueducts in the mid-first century, it had no doubt become an important but secondary aqueduct within the overall system, used primarily for industrial purposes.

The Anio Vetus therefore played several roles in its history. As the second aqueduct in Rome's water system, it marked a major step forward from the Appia. The evidence suggests, however, that it was originally planned to complement the earlier line, to increase the supply of water available in the center of Rome, and to furnish drinking water to those higher areas in the eastern part of the city that the Appia could not serve. It was therefore brought to the Esquiline, from where it was distributed further to the Viminal and perhaps to the Quirinal, first within the circuit of the "Servian" Wall, then beyond its perimeter as the city expanded in the late Republic. Its introduction made possible supply to the Subura and that part of Region VIII (Forum Romanum) not adequately served by the Appia. Because of its high point of entry, its water, although carried at a level lower than that of later aqueducts, was subsequently extended as far as the Campus Martius and Transtiber, and the branch of the *specus Octavianus* brought water to the area of the Little Aventine, probably in the Augustan period.

The capacity of the Anio Vetus, much greater than that of the Appia, made it one of the four major aqueducts of Rome, with a volume second only to that of the Virgo and the combined Aqua Claudia and Anio Novus.[21] Yet by Frontinus' time its water was considered inferior to others. This might explain its heavy distribution for private use and the large amounts Frontinus reports were illegally diverted from it (66.7); such diversions would mainly have been at the expense of private consumers, rather than public needs.

---

vicinity of the Porta Esquilina was reworked, at least in part, in the Trajanic period. See Santa Maria Scrinari, 61–62. In addition, major improvements were made by later emperors on the conduit outside the city; the Hadrianic "Ponte della Mola" near San Gregorio is perhaps the most dramatic example of such upgrading. See Ashby, 68–70; Van Deman, 63; Quilici, "Acquedotti," 65–66.

21. Cf. Blackman, "Volume."

CHAPTER 5

The Aqua Marcia

A period of almost 130 years separates the Anio Vetus from the introduction of Rome's third aqueduct in 144 B.C. But the increased building activity in the public sphere attested by our literary sources points to the steady growth of the city and its population,[1] especially in the years after the Second Punic War and during Rome's expansion into the East in the second century B.C. As a natural consequence, there was a dramatic increase in its need for water; Frontinus himself cites the *incrementum urbis* in 7.2.[2] Accounts of the censorship of Cato the Elder in 184 B.C. include notices that the censors reclaimed public water flowing onto private property and also let contracts for the paving of *lacus,* both actions indicating a concern to make best use of what must have been considered a limited resource (Livy 39.44.4–5; Plut. *Cat. Mai.* 19).

Also significant is Livy's notice (40.51.7) that the censors of 179 B.C., Aemilius Lepidus and Fulvius Nobilior, let contracts for the introduction of an aqueduct and construction of arches (*aquam adducendam fornicesque faciendos*); the project was, however, blocked by M. Licinius Crassus, who refused them a right-of-way through his property. Livy's detail about arches is worth noting: since Rome's two earlier aqueducts ran almost completely underground, the censors of 179 must have intended to introduce a comparatively high line, a project much more ambitious than that of the Appia or Anio Vetus.[3] After the defeat of Hannibal,

---

Sources: *Aq.* 7; 12.1–3; 18.3–4; 19.1–6, 8–9; 67; 76.5; 81; 91.5

1. For a catalog after 200 B.C. see F. Coarelli, "Public Building in Rome Between the Second Punic War and Sulla," *PBSR* 45 (1977): 1–19.

2. For a review of the situation and growth of the city see M.G. Morgan, "The Introduction of the Aqua Marcia into Rome, 144–40 B.C.," *Philologus* 122 (1978): 25–28. Cf. also Brunt, 384.

3. See M. Stuart, "The Denarius of M'. Aemilius Lepidus and the Aqua Marcia," *AJA* 49 (1945): 242.

military considerations would no longer have dictated underground conduits (18.4), and there was no doubt a feeling that water now had to be brought in at a level high enough to supply much wider and higher areas of the city.

We hear nothing more about the aqueduct of 179 B.C., or how much of it, if any, was built.[4] Nor is there evidence of further efforts to increase the water supply during the next thirty-five years. However, things may well have reached a crisis in 144, when Q. Marcius Rex was assigned the task of restoring the two existing aqueducts and bringing in a new supply of water.[5] Marcius' commission, as reported by Frontinus (7.2), did not necessarily stipulate the introduction of a new aqueduct (*quatinus alias aquas quas posset in urbem perduceret*), but that is what he did. The result was a line much more ambitious in its technical aspects, drawing its water from a source more distant than that of the Anio Vetus, with a conduit of 61,710½ *passus*, of which 7,463 *passus* were above ground, 528 on substructure, and 6,935 on arches (7.8.) The cost of the construction, certainly very great (7.4), must have been financed from the booty taken from Corinth and Carthage after 146 B.C.[6] Indeed the sudden influx of cash into Rome at this time perhaps made possible the completion of a project which had been planned for some time.

The circumstances of the construction, as reported by our sources, were unusual. Marcius is the only urban praetor recorded as the builder of an aqueduct, and the Marcia was the only Republican line built following a *senatus consultum*.[7] Moreover, the speed with which the aqueduct was completed is striking, given its length and height.[8] The Marcia's branch to the Capitol was famous in antiquity and commemorated by an honorary statue of its builder located, perhaps near the terminus, *post aedem Iovis Optimi Maximi*, as recorded by a military diploma found in Bavaria (*CIL* 3.846=16.5).[9] The oppo-

---

4. Morgan, 36–37, concludes that the line of 179 B.C. was never begun. Cf. Stuart, 243–46, who argues that Marcius exploited the arches of the earlier line begun by the Aemilii Lepidi to complete his aqueduct so speedily and posits a name "Aemilia-Fulvia" for the aqueduct of 179. His arguments are rejected by Morgan, 35–43; and Coarelli, *Foro Boario*, 143–44.

5. For a full discussion of the circumstances see Morgan, 48–54.

6. T. Frank, *An Economic Survey of Ancient Rome*, vol. 1: *Rome and Italy of the Republic* (Baltimore, 1933), 226–29, estimates that the amount, 180 million sesterces, was equal to four and a half years' revenue. Cf. Coarelli, "Public Building," 17.

7. Rodgers in his forthcoming commentary on 7.2 observes that a public emergency may well have existed if problems of water supply could not wait until the appointment of new censors in 142 B.C.

8. Cf. Stuart, 243–46; M. Hainzmann, "Zur kapitolischen Wasserleitungen des Q. Marcius Rex," *GrazBeitr* 6 (1977): 31. Robinson, "Water Supply," 49, describes the Marcia as the first aqueduct in Rome to have "an obvious, doubtless dramatic appearance."

9. Lanciani, 311; Hainzmann, *Untersuchungen*, 91 n. 4.

sition to its extension, the interpretation of which is greatly complicated by textual difficulties in Frontinus' account (7.5), has also been widely discussed. From the evidence M. Gwyn Morgan has recently argued that the *quindecimviri* opposed the introduction of any aqueduct to the Capitoline, not specifically the Aqua Marcia itself, to prevent settlement of *loca publica* under their control.[10] More important for our discussion is the aqueduct's point of entry into the city, which determined its subsequent distribution.

Like the two earlier aqueducts, the Marcia approached Rome at Spes Vetus, where remains of its conduit, carrying those of the later Tepula and Julia, can still be seen incorporated into the Aurelian Wall (figs. 2–3).[11] From this point it continued northwest, crossing the Via Tiburtina, where its conduit is preserved in the Augustan Porta Tiburtina (fig. 4),[12] then disappeared along the present Via Marsala until it reached the Porta Viminalis, the highest point within the circuit of the "Servian" Wall, where it turned southwest to enter the city. Its course, outlined by Frontinus (19.3–5), can also be traced through remains of the channel and *cippi* along the route (fig. 1).[13]

The archaeological evidence indicates that widespread distribution began in the area of the Porta Viminalis. However, the only relic of this distribution visible until quite recently was a small circular structure of tufa and travertine outside the line of the "Servian" Wall in the present Piazza dei Cinquecento (fig. 5).[14] The structure obviously served distribution purposes, since lead pipes issued from two holes on its north side. But because it stands above conduit level, its exact function is unknown: water would have to have been forced up into it, perhaps by the use of a standpipe like those found at Pompeii. Its position outside the Porta Viminalis and small size suggest that it was part of only a secondary branch.[15]

The location of the main *castellum* inside the Porta Viminalis is unknown, since the *cippi* recording the route of the conduit found within the wall were probably not in situ. One branch, however, ran northwest from the gate toward the area presently occupied by the Ministry of Finance, where traces of three

10. Morgan, 48–54.
11. Nash, fig. 43.
12. Nash, figs. 44–45.
13. D. Cattalini, "Aqua Marcia," in *Trionfo*, 49–50; *CAR*, 3.271–72, no. 218; Ashby, 146–49; Van Deman, 116–22; Lanciani, 303–5; Hainzmann, *Untersuchungen*, 96–97.
14. The structure has now been dismantled for subway construction. See Nash, fig. 46; *CAR*, 3.254, no. 167–68 III f; and Ashby, 149, for a full description.
15. On standpipes in Pompeii see Hodge, 300–303. Panimolle, 93, states that it is the terminal *castellum* for the Marcia, Tepula, and Julia, an impossible identification, given its limited size. Cf. Hainzmann, *Untersuchungen*, 98 n. 2.

castella were found during its construction. Perhaps these are to be connected with the Marcia, Tepula, and Julia.[16] This position would have been well chosen, since from here a secondary line or lines could have been easily run along the ridge of the Quirinal to supply that hill, the lower ground of the city center, around the Forum Romanum,[17] and extended to the Capitoline. The Quirinal branch could have run atop the "Servian" Wall, along the northern ridge of the hill. Its use of the wall for delivery in that district may well explain the relatively good preservation of long stretches of the wall's circuit in that area, in contrast to other districts of the city; if the "Servian" Wall along the Quirinal did carry a branch of the Marcia, it would have been maintained long after it ceased to serve any defensive function.[18] No traces remain of the famous Capitoline branch, which probably delivered a relatively small supply by means of a siphon over the natural spur of high ground that connected the Capitoline to the Quirinal before construction of the Forum of Trajan.[19] Another subsidiary line has been identified with remains of a channel found in the area of the present Via Cattaneo and Piazza Fanti.[20]

---

16. Lanciani, 306–7; Ashby, 151.

17. A. Scobie, "Slums, Sanitation and Mortality in the Roman World," *Klio* 68 (1986): 423, states that the Lacus Servilius near the Temple of Saturn was fed by a branch of the Marcia; this may be correct, since the fountain dates to the Republican period, but we have no evidence that the Marcia itself supplied it. For the Lacus Servilius, later restored by Agrippa, see Platner-Ashby, 314; Richardson, 232; and Roddaz, 150.

18. On city walls supporting water conduits see Hodge, 291. Buzzetti, 24, suggests that it was "not improbable" that the Marcia used stretches of the "Servian" Wall for distribution within the city, but cites only the "Arch of Dolabella" on the Caelian, the "Arch of Gallienus" on the Esquiline, and the "Arch of Lentulus" in the Forum Boarium, not the wall along the Quirinal. For its preservation in this area see G. Säflund, *Le mura di Roma repubblicana* (Uppsala, 1932), 76–98.

A topographical reference in Mart. 4.18, which describes the death of a boy from a falling icicle as he passed under a *vicina . . . Vipsanis porta columnis* (4.18.1), is commonly associated with the line of the Aqua Virgo near the Porticus Vipsania in the Campus Martius; cf., e.g., L. Friedlander, ed., *M. Valerii Martialis Epigrammaton Libri*, 2 vols. (Leipzig, 1886), 1:344; and, most recently, Rodriguez Almeida, in *FUM*, 123 n. 7; and Roddaz, 292. The detail about the *porta* may, however, refer to a gate of the "Servian" Wall in proximity to Agrippa's complex: a leak from the line of the Marcia running atop the wall may have produced the fatal icicle. Martial, who lived on the Quirinal (5.22.3, 6.27.1, 10.58.9, 11.1.19), also cites the Marcia in another poem (9.18.6).

19. See A.W. Van Buren, "Come fu condotta l'acqua al Monte Capitolio?" *RendPontAcc* 18 (1941–42): 65–70; and G. de Angelis D'Ossat, "La sella fra il Campidoglio e il Quirinale: Acquedotto Marcio 'in Capitolium'," *Capitolium* 21 (1946): 17–23. Cf. also Bruun, 134; Lanciani, 311; Ashby, 152; Van Deman, 139. Cic. *Pro Rabirio Perduellionis Reo* 11.31 cites *fistulae*, probably of this branch. On Roman use of siphons see Hodge, 147–60.

20. Ashby, 150.

Frontinus specifically mentions one branch line, the *rivus Herculaneus*, probably because unlike the others, it deviated from the main conduit in the vicinity of Spes Vetus (*post hortos Pallantianos*), before the Marcia reached its terminal *castellum* (19.8–9). His descriptions of its course under the Caelian and its terminus *supra portam Capenam* have in part been confirmed by archaeological evidence: Piranesi drew a plan of the *castellum* at its point of divergence, a structure now no longer visible, which was incorporated into the line of the Aurelian Wall,[21] and traces of the conduit have been found in various areas of the Caelian, paralleling the line of the later Neronian *Arcus Caelimontani*.[22] However, the *rivus Herculaneus* remains a problematic branch, because Frontinus gives no information concerning its date of construction or the purpose it served. Its name is perhaps to be associated with the temple (or statue?) of Hercules Sullanus listed in the Regionary Catalogs,[23] but its course is unusual: since Frontinus reports that it was too low to supply the Caelian (*ipsius montis nihil ut inferior subministrans*, 19.9), its distribution could have been carried out only through pipes running to its slopes, an arrangement impractical for delivery to the hill itself. The term *rivus* itself may also be significant here; because it is not designated a *specus*, the branch may well have been, at least in large part, a cut-and-cover channel running along the shoulders of the hill, rather than through a tunnel underneath it.

The terminus of the *rivus*, at a point *supra portam Capenam*, suggests that it was introduced to serve needs of the immediate district no longer adequately supplied by the nearby spring of the Camenae.[24] Because Frontinus indicates that the branch ended *supra portam Capenam* and did not cross to the Aventine, its terminal has been identified with the famous *arcus stillans* mentioned by the scholiast *ad* Juv. 3.11: *veteres arcus madidamque Capenam. quia supra eam aquaeductus est, quem nunc adpellant arcum stillantem. primum enim usque ibidem fuerunt portae* ("old arches and the wet [porta] Capena: since there was an aqueduct above the gate, which they now call the dripping arch. Indeed there first were gates in the same place").[25]

The passage raises a long-standing topographical problem. Jordan interpreted

21. Van Deman, 140, textcut 14.

22. Ashby, 154–55; Van Deman, 139–41; Colini, *Topografia*, 82–84, 88; Hainzmann, *Untersuchungen*, 98–99.

23. Nordh, 80; Platner-Ashby, 256; Richardson, 188. Grimal, 13 n. 3, suggests that the name was a common one, comparing the *rivus Herculaneus* outside the city, which Frontinus cites in 15.4.

24. Lanciani, 223–25; Platner-Ashby, 89–90; Colini, *Topografia*, 12–14.

25. Ashby, 155; Van Deman, 141. Cf. Mart. 3.47.1: *Capena grandi porta qua pluit gutta*.

the *arcus* as a nickname for the Porta Capena itself, wet because of leaks from a nearby aqueduct.[26] But the evidence of the twelfth-century *Mirabilia Urbis Romae* suggests that the *arcus* was a separate structure inside the gate, not the gate itself (*intus portam arcus stillae,* 10.2).[27] Since we would expect any aqueduct in the area to have run inside the circuit of the "Servian" Wall, rather than outside it, the *arcus stillans* may have had nothing to do with the *rivus Herculaneus* but is perhaps rather to be identified with part of the arcuation of the Appia that crossed the valley between the Caelian and Aventine (5.5).[28]

Also problematic is another notice in Frontinus that the Caelian and Aventine, before construction of the Aqua Claudia, were supplied by the Marcia and Julia (76.5). Since the *rivus Herculaneus* was too low to furnish water to the Caelian (19.9), this can only mean that another conduit at a higher level distributed to these hills. No trace of any earlier line of the Marcia has been identified, and the evidence for this high-level distribution, which we would expect to have run above ground (at least in the neighborhood of Spes Vetus), is scanty indeed: a fragment of a marble inscription, now lost (*CIL* 6.1247), analogous to the Severan inscription on the Porta Tiburtina (*CIL* 6.1245), appears to record the repair of a branch of the Marcia in A.D. 196 and is perhaps to be connected with a monumental arch carrying the line over a principal street.[29] Attempts have also been made to associate the "Arch of Dolabella" with this high-level line.[30] Yet the arch, which is not freestanding but bound on its north side to a wall of grotta oscura tufa, is rather to be connected with the line of the "Servian" Wall.[31]

---

26. Jordan, 2:19.

27. Jordan, 2:615. For a different location of the *arcus stillans* cf. Richardson, 92, arguing for its identification with the mysterious Porta Romana (or Romanula) cited in Festus 318L.

28. So Panimolle, 44–45, but for the implausible reason that the Appia as the oldest line would be the most subject to leaks. This arcade of the Appia is perhaps also to be identified with the mysterious Duodecim Portae of the Regionary Catalogs (Nordh, 91). See Richardson, 144; Jordan, 2:88; and Platner-Ashby, 198. Pace, 253 n. 6, identifies the leak with the *rivus Herculaneus,* which he suggests terminated at the Thermae Suranae on the Aventine. Buzzetti, 24, associates it with the Marcia, which he suggests was carried on the "Servian" Wall in this area. Lugli, 2.355, and Panimolle, 95, argue that the *rivus* was abandoned after introduction of the Aqua Claudia into the area.

29. Colini, *Topografia,* 85; the inscription is cited by Ashby, 90–91, but in connection with the Marcia itself, not with this branch.

30. Ashby, 155; Van Deman, 142; Lanciani, 312–13; Panimolle, 95–96; Hainzmann, *Untersuchungen,* 99 n. 3.

31. Colini, *Topografia,* 33–35; Coarelli, *Foro Boario,* 45–47. The fragment of an Augustan *cippus* (*CIL* 6.31510) cited by Van Deman, 142, gives no conclusive evidence

The only other evidence for this high-level branch of the Marcia is a second notice by Frontinus that substantial reconstruction by Trajan brought the Marcia's water to the Caelian and Aventine (*et in primis Marcia reddita ampliore opere a Spe in Aventinum usque perducitur,* 87.4). Interpretation of this passage is difficult, because *reddita* does not necessarily imply that the Marcia had been distributed there previously.[32] Frontinus' use of the participle does, however, suggest that Trajan restored a distribution system previously in operation.

This Trajanic restoration of the Marcia is itself also problematic. All physical evidence of it has disappeared, although attempts have been made to identify it with scattered remains, a *specus* found on the Aventine northwest of the church of S. Prisca, and sixteen arches recorded in 1643 northwest of the church of S. Balbina.[33] We must, however, accept Frontinus' notice about Trajan's construction if we are to make sense of his observation following immediately in the same chapter (87.5), that all points of the city now enjoyed a dual distribution of water so that supplies could be maintained even if one line was interrupted.[34]

Frontinus records two earlier restorations of the Marcia, the first by Agrippa in 33 B.C. (9.9) and a second by Augustus, who significantly increased its capacity by adding a supplemental source known as the *aqua Augusta* (12). At a later period, after introduction of the Aqua Claudia lessened demand on the Marcia itself, the *aqua Augusta* seems to have become primarily a reserve supply used to augment other lines as needed, particularly the Claudia (14.3, 72.8). Augustus' overhauling of the Marcia is also attested by his inscription on the Porta Tiburtina (*CIL* 6.1244), and its inclusion among his achievements in the *Res Gestae* (20.2) is proof of the importance he attached to it. We also know of later restorations: one by Titus in A.D. 79 (*CIL* 6.1246), perhaps to prevent contamination of the source by flooding in the Arsoli basin,[35] and that by Septimius Severus in A.D. 196 already cited (*CIL* 6.1247), which might be associated with the increased demand caused by the Severan Baths in Region I.[36] Caracalla's restoration of the Marcia in A.D. 212 (*CIL* 6.1245) involved the addition of a new source, the *fons novus Antoninianus,* to compensate for the loss of water

---

about the line. Cf. Buzzetti, 24, who suggests that the wall may have carried an aqueduct at this point, probably that of the Marcia.

32. Cf. Colini, *Topografia,* 87: "Evidentemente la Marcia era già distribuita al Celio, se poté essere *reddita.*"

33. Ashby, 156.

34. Cattalini, "Aqua Marcia," 50; Hainzmann, *Untersuchungen,* 100 n. 1.

35. Lanciani, 283–84.

36. Lanciani, 272–73.

diverted for his bath complex on a special branch, the *aqua Antoniniana*.[37] There were further repairs by Diocletian, connected with the construction of his bath complex,[38] and restorations attested as late as Arcadius and Honorius.[39]

From the large number of its branch lines already mentioned and its much higher point of entry into the city, we should expect a much wider distribution of the Marcia than that of the two lines that preceded it. Indeed Frontinus lists the Marcia as one of the five aqueducts whose height allowed them to be supplied to every part of the city (18.3–4). The distribution scheme that he gives elsewhere (81.2) reflects the hydraulic plan of the city when he entered office (*copia aquarum ad Nervam imperatorem usque conputata*, 87.1) before completion of Nerva's *amplius opus* (87.4), since it excludes the Caelian (II) and Aventine (XIII), both hills to which he specifically states the Marcia had recently been brought. However, Frontinus also indicates a widespread distribution even before Nerva's work: after its delivery to the Caelian and Aventine, the aqueduct supplied every region of the city except the Circus Maximus (XI) and Piscina Publica (XII).

It is worth noting that the Marcia supplied the Palatine (X), the first line to be brought to that hill. No doubt its higher level made such delivery possible, probably by means of a siphon.[40] Introduction of the Marcia also made possible increased distribution to the city center, the Forum Romanum, and the southern Campus Martius, as well as augmented delivery to eastern hills of the city, at a level higher than that of the Anio Vetus.[41] The number of its *castella* that Frontinus lists, 51 in all (81.2)—much greater than the 20 of the Appia (79.2) and 35 of the Anio Vetus (80.2) and one-fifth of the total of 247 within the city at the end of the first century A.D. (78.3)—also indicates widespread distribution through multiple secondary lines.

This greater number of *castella* also points to a notable increase in capacity over that of earlier lines. Here, however, our knowledge is limited because of wide discrepancies in the statistics given by Frontinus' text. In his chapter discussing his measurement of its volume Frontinus cites a capacity of 4,690 *quinariae*, 2,528 more than in the record books (67.1–2).[42] He also lists two

---

37. Cattalini, "Aqua Marcia," 57–59; Ashby, 91; Van Deman, 68; Lugli, 2:356; Lanciani, 315–18; Hainzmann, *Untersuchungen*, 102.

38. Ashby, 151.

39. Ashby 92–93; Lanciani 284–86.

40. Bruun, 134.

41. Hainzmann, *Untersuchungen*, 104.

42. This discrepancy is striking, since the Marcia's actual capacity would have been almost double that listed in the records. Could the official *commentarii* have reflected only figures prior to Augustus' supplement to the line, through which he claimed to

supplements given to other lines by the Marcia, 92 *quinariae* to the Aqua Tepula and 164 to the Anio Vetus (67.3),[43] and gives a total delivery of 1,840 *quinariae* below its *piscina* (67.7). But in his analysis of its delivery (81) he lists distribution of 261½ *quinariae* outside the city and 1,472 *quinariae* inside, or a total of 1,733½ *quinariae*. We therefore find a difference of 106 *quinariae* between the two chapters, no doubt the result of the faulty text.

Even more serious, however, is the disagreement of the arithmetic within chapter 81 itself, complicated by the loss of one line in the MSS. giving the total number of *quinariae* supplied for *usus publici*. Frontinus reports a total delivery of 1,472 *quinariae*, but addition of the separate volumes listed in the lines that follow immediately results in a much lower figure:

Marcia distribution	1) *nomine Caesaris*	116 q.	(10.5%)
intra urbem	2) *privatis*	543 q.	(49.3%)
	<3) *usibus publicis*	443½ q.	(40.2%):>
	a) 4 *castra*	42½ q.	(3.9%)
	b) 15 *opera publica*	41 q.	(3.7%)
	c) 12 *munera*	104 q.	(9.4%)
	d) 113 *lacus*	256 q.	(23.2%)
	TOTAL	1,102½ q.	

To make sense of the manuscripts readings we must either emend the text radically or accept one set of statistics over another. The total of 1,472 given first is closer to the figure reported earlier in chapter 67. But those provided by Frontinus' analysis of the distribution carry more weight. Not only are they more detailed than the single total, but the statistics that they give for public basins also give them greater credibility: delivery of 256 *quinariae* to 113 *lacus* results in an average of 2.26 *quinariae* per basin, a figure analogous to those reported for the Appia and Anio Vetus.[44] Our analysis therefore follows the totals reflected by the detailed analysis of distribution, and the percentages given above are based on a total of 1,102½ *quinariae*. We should, however, keep in mind that the statistics are uncertain and quite likely do not accurately reflect the total distribution of the Marcia at Frontinus' time: a discrepancy as large as 369½

---

have doubled its capacity (*Res Gestae* 20.2)? As in the case of the Appia, we would expect the records to have been brought up to date. In addition, Frontinus asserts that he discovered 2,499 *quinariae* being illegally diverted (67.8).

43. Grimal, 33 n. 2, noting these volumes given to other lines, suggests that the Marcia, because of its volume, was a convenient backup to cover shortages in other aqueducts from fraudulent deliveries.

44. Lanciani, 579; Bruun, 102.

quinariae results in a margin of error of some 25%, perhaps too high to permit significant conclusions from our evidence.

Several observations, however, can be made about these figures, problematic as they are. Very little of the Marcia's water was delivered *nomine Caesaris,* only some 10.5% of its total distribution, and relatively little of it went to *opera publica* and *munera:* the aqueduct supplied only 3.7% of its water to public buildings and 9.4% to ornamental fountains. These figures at first seem surprisingly low, given its height at its entry into the city and wide distribution throughout Rome. Its supply to *castra* is also small, reflecting a percentage similar to that of the Anio Vetus. What is most striking about the Marcia's distribution is the high percentage of its water delivered to *privati* and to public *lacus,* almost a half and a quarter, respectively, of the total supply (49.3% to *privati* and 23.2% to basins). Nearly three-fourths of the Aqua Marcia was delivered for drinking, either to private citizens or through public basins in Rome, some 72% in all.

Frontinus' data therefore confirm the widespread testimony about the quality of the Marcia's water (*Marciam ipsam frigore et splendore gratissimam,* 91.5; cf. Pliny *HN* 31.41: *clarissima aquarum omnium in toto orbe frigoris salubritatis palma;* Mart. 8.16.6). For this reason Frontinus cites his efforts to preserve the integrity of the line, keeping it for human consumption whenever possible within the city (92).[45] That the Marcia continued to serve this primary purpose even after his time is evident from what we know about later restorations: when Caracalla diverted part of its water in A.D. 212–13 to supply his bath complex, he added a supplemental source, the *fons Antoninianus,* to compensate for the resulting loss (*CIL* 6.1245). In like manner Diocletian constructed an additional channel, the *forma Iovia,* in connection with deliveries to his Thermae.[46]

All our evidence therefore indicates that the Marcia's water was considered special, delivered as widely as possible within the city, and reserved for drinking rather than supplying public works or other needs. When it was later tapped for imperial bath complexes, additional waters were secured to offset those taken for that delivery. Although water of other lines, particularly that of the Aqua Claudia (89.4) and even the Anio Novus (94.3), was of comparable quality at their sources, these aqueducts do not seem to have been as highly regarded in Rome. Introduced originally to meet the growing needs of the city in the

---

45. Grimal, 31 n. 1, suggests that the limited delivery of the Marcia *extra urbem* before reaching its *piscina* (95 *quinariae* in all, 67.3), indicates a deliberate plan to reserve as much of it as possible for drinking within Rome.

46. Ashby, 14, 91–92, 151–52, 156–58; Lanciani, 318–19.

second century B.C., the Marcia became relatively insignificant in its overall capacity, especially in comparison with the volume of later imperial aqueducts, but it was always regarded as premium water, with a reputation like that enjoyed until recently by the modern Marcia-Pia.[47]

47. For the modern Marcia-Pia see A. Mastrobuono, "L'acquedotto Marcio," in *Trionfo*, 287–95.

CHAPTER 6

The Aqua Tepula

Of all the Republican aqueducts the Aqua Tepula is the most problematic, since we know almost nothing about its original purpose; nearly all the information given by Frontinus refers only to the Tepula of the first century, after reworking by Marcus Agrippa had radically changed its original character. Constructed only nineteen years after the Marcia, the line seems to have been almost an afterthought, a much shorter aqueduct of limited capacity probably intended as a supplement to the Marcia.

Frontinus' notice of its introduction in 125 B.C. is itself unusual (8.1–2): the censors responsible contracted for bringing water to Rome and the Capitol (*Romam et in Capitolium adducendam curaverunt*), thus specifying an area of the city that had been supplied by aqueduct for the first time only fifteen years earlier through the famous branch of the Marcia. The introduction of a second aqueduct within twenty years indicates a steady growth in the city's population,[1] and the detail about the Capitol may point to an urgent need that the branch of the Marcia, no doubt of limited capacity, could not supply. It is also worth noting that in contrast to the circumstances surrounding the Marcia's earlier extension to the hill, Frontinus records no opposition to the project.[2]

No traces of the original Tepula have been found, and it has therefore been suggested that its earliest channel, quite different from the Agrippan construction that survives today, was abandoned after its reworking in the late first century B.C.[3] But here Frontinus' notice of distribution to the Capitol is significant: to

Sources: *Aq*. 8; 18.4; 19.1–6; 68; 82

1. Brunt, 384.
2. Cf. Morgan, 53.
3. Van Deman, 147; F.W. Shipley, *Agrippa's Building Activities in Rome* (St. Louis, 1933), 28–29; Panimolle, 101.

be able to bring water to that hill, the original Aqua Tepula would have to have been a high-level line, comparable to the Marcia. But the aqueduct was of very limited capacity. Frontinus states that it had no source (*fontes*) of its own but was drawn from rivulets (*venae*) from the springs that later supplied the Julia (68.2–3). It would therefore seem highly improbable that the censors of 125 B.C. would have contracted for construction of a second, high-level line separate from the Marcia to furnish only a limited amount of water to the city.[4] The small increase that it would have added to the overall supply could hardly have justified the expense, even though the water was brought from a relatively short distance. Instead we must conclude that from the beginning the Aqua Tepula, after running from its source in the Alban Hills to the modern Capannelle,[5] used the arches of the Marcia from that point for its entry into Rome.

In comparison with the Marcia of nineteen years earlier, the Tepula was therefore a much less ambitious project. Its limited length and capacity were perhaps dictated by economic considerations during the politically unstable decade of the 120s B.C. Because it came into Rome on the arches of the Marcia (19.1–4), its level was slightly higher than that of the earlier aqueduct (18.4).[6] But in all other respects it was much less impressive. As its name indicates, the tepid quality of its water probably made it unpalatable, a shortcoming even more striking in comparison with the excellence of the Marcia, which had immediately preceded it.[7] From the very beginning its use must have been restricted within Rome, and it may well have been introduced specifically for industrial purposes, to permit Marcian water to be reserved for drinking.

Frontinus' data concerning the Tepula are applicable only after Agrippa's extensive reworking in 33 B.C. To lower the temperature of the water and augment volume, Agrippa added to it water from his new line, the Aqua Julia, together with which the Tepula was carried to its *piscina* at the sixth milestone on the Via Latina. Here the waters of the two aqueducts were again separated into different channels, and the Tepula rode into Rome under its own name (19.3). But separation of the Tepula and Julia into two *specus* was now only a nominal division, since the waters of the two lines had been mixed. After Agrippa, the

4. Cf. Quilici, "Acquedotti," 74, who suggests that the original Tepula could have run on the arches of the Marcia but that its conduit of the second century was demolished by Agrippa and rebuilt to support the weight of the Aqua Julia atop it.

5. For the location of the source see D. Cattalini, "Aqua Tepula," in *Trionfo*, 60–61; and Ashby, 159–60.

6. The most recent levels taken by the Cooperativa MODUS at the Porta Maggiore indicate a level of 57.61 masl. for the Tepula and 55.70 masl. for the Marcia. See G. Tedeschi Grisanti, "Primo contributo ad una livellazione urbana sistematica degli antichi acquedotti di Roma," in *Atti del convegno "Gli antichi acquedotti di Roma: problemi di conoscenza, conservazione e tutela"* (Rome, 1992), 59–69.

7. Ashby, 159–60, gives the temperature of its source as 61–63 degrees F. in winter.

Tepula effectively ceased to have an identity of its own. Its total volume consisted of water received from the Julia (190 *quinariae*), with supplements of 92 *quinariae* from the Marcia at its *piscina* (67.3, 68.3) and 163 *quinariae* near the Horti Epaphroditani from the Anio Novus (68.4). By Frontinus' time the aqueduct had therefore become a composite of various lines delivered through a common distribution system.[8]

From archaeological evidence we can trace the route of the Tepula into the city, either atop the Marcia's arcade or parallel to its *specus* underground from the Porta Maggiore to the Porta Viminalis (figs. 1–3). At this point, like the Marcia, it turned southwest to enter the circuit of the "Servian" Wall.[9] Its terminal *castellum*, like that of the Marcia, is perhaps to be connected with remains found under the Ministry of Finance, northwest of the Porta Viminalis, but nothing further is known about its distribution. The Capitoline branch, like that of the Marcia, probably ran along the ridge of the Quirinal and crossed to the Capitoline Hill by a siphon.

It should be noted that the Tepula, alone of all the aqueducts listed by Frontinus, lost none of its *quinariae* between its source and terminus (68.5).[10] Moreover, in contrast to his reports on other lines, Frontinus gives no length for the Tepula, nor does he state that he measured its volume, relying instead on the figures in the record books (68.1). The aqueduct was clearly of secondary importance to him. Its capacity was very limited in comparison with that of other lines, especially when we take into account the 225 *quinariae* it received from the Marcia and Anio Novus.

The Tepula delivered only 331 *quinariae* through 14 *castella* to four regions, Templum Pacis (IV), Esquiliae (V), Alta Semita (VI), and Via Lata (VII), all of which were also served by the Marcia (82.2). What is surprising about Frontinus' notice is the omission of Region VIII, in which the Capitoline is located. This may merely reflect an error in the MS. tradition, but it may also indicate that by the end of the first century A.D. the Tepula no longer supplied the area for which it was originally introduced (8.1). Since the Aqua Julia, a line of greater capacity that entered Rome at a slightly higher level (19.4), is listed as distributing to Region VIII (83.2), its water may have been substituted for that of the Tepula at a later period.[11]

---

8. Frontinus gives no statistics on the length of the Tepula in his survey of individual lines (8.1–2), probably because he did not consider it an independent part of the water system.

9. Cattalini, "Aqua Tepula," 60–61; *CAR*, 3.272, no. 218; Ashby, 147–51; Van Deman, 148.

10. Hainzmann, *Untersuchungen*, 108.

11. Van Buren, "L'Acqua al Capitolio," 65, suggests that this omission indicates a

Frontinus gives the following analysis of its distribution (82.2):

Aqua Tepula distribution *intra urbem*	1) *nomine Caesaris*	44 q.	(13.3%)[12]
	2) *privatis*	237 q.	(71.6%)
	3) *usibus publicis*	50 q.	(15.1%):[13]
	a) 1 *castra*	12 q.	(3.6%)
	b) 3 *opera publica*	7 q.	(2.1%)
	c) 13 *lacus*	32 q.	(9.6%)
	TOTAL	331 q.	

Almost three-fourths of the Tepula was delivered to private citizens and relatively little for imperial buildings or public uses, including less than 10% for public basins. It is also to be noted that the Tepula, alone of all the aqueducts discussed so far, supplied no *munus*. The evidence therefore suggests that by Frontinus' time it functioned mainly as a supplemental line serving the eastern districts of the city that were also supplied by other aqueducts. Although it entered the city at a level high enough to permit widespread distribution (18.3–4), its limited capacity did not make such delivery feasible. Instead it appears to have been a secondary line, supplying water mainly to private consumers in districts to which it could be readily be brought from its point of entry at the Porta Viminalis. Its limited distribution may well have been tied to a poor reputation; even after Agrippa had improved its quality, the original name remained.[14]

---

modification in the urban distribution system between the second century B.C. and the time Frontinus wrote his treatise.

12. Reading with Buecheler 44 *quinariae,* for the 34 of the MSS., to make sense of the arithmetic. The discrepancy does not affect conclusions to be drawn from the data.

13. Lanciani, 313–14, suggests that Frontinus' calculations are off by one *quinaria* because he rounded off fractions.

14. Rodgers in his commentary on 8.1 suggests that the name Tepula may indicate a primitive cult associated with the water of the source; in this way the name would have been preserved long after the temperature of its water was lowered.

CHAPTER 7

The Aqua Julia

Although Rome had grown steadily during the period following the introduction of the Tepula in 125 B.C., the political chaos of the last century of the Republic and resulting civil wars no doubt precluded the construction of any major new aqueduct. Indeed the evidence suggests that by the time Agrippa assumed the aedileship in 33 B.C.,[1] Rome's water system was in urgent need of a general overhauling. Not only did Agrippa repair the conduits of the first three aqueducts, which Frontinus describes as almost totally worn out at that period (*paene dilapsos*, 9.3), but, even more important, he also put the administration of the water supply on a sound footing for the first time. After his term as aedile, acting as a quasi-permanent water commissioner (*velut perpetuus curator*, 98.1), Agrippa fixed a distribution plan for the water then available and established a permanent staff for its operation and maintenance. After his death, these innovations became the basis for the *cura aquarum* established in 11 B.C. (99.1–5).[2]

In addition, Agrippa introduced a fifth line into the city. Having tapped new springs near the source of the Tepula in the Alban Hills, he built another

---

Sources: *Aq.* 9; 18.4; 19.1–7; 69; 76.5–7; 83

1. The dating of the Aqua Julia is disputed. A notice in Dio Cass. 49.32.3–4 states that the line was introduced in 40 B.C. at the end of Agrippa's censorship. Ashby, 161 n. 5, favors the earlier date, suggesting that the Julia may have been Julius Caesar's project, which Agrippa completed after the dictator's death. This might also account for the name, usually explained as an example of Agrippan self-effacement, like his naming of the Portus Julius near Misenum. Cf. M. Reinhold, *Marcus Agrippa: A Biography* (Geneva, N.Y., 1933; repr. Rome, 1965), 49. But 33 B.C. would appear to be a more likely date, given the political situation at the time. Cf. Roddaz, 149; Hainzmann, *Untersuchungen*, 108 n. 1; and Shipley, 19–20, 27–28.

2. Shipley, 24–26; Bruun, 148–49; W. Eck, "Organisation und Administration der Wasserversorgung Roms," in *Wasserversorgung*, 66–71.

aqueduct, the Julia, which furnished additional water to the Tepula and brought a much larger supply of its own into Rome (9.1–3, 19.1–3). After his construction, the waters of both the Tepula and Julia, having been mixed at their *piscina* outside the city, were carried into Rome in separate conduits on the arches of the Aqua Marcia. The route of the Julia, like that of the two earlier lines, can therefore be traced from Spes Vetus to the Porta Tiburtina, then underground to the Porta Viminalis where it turned to enter the city (figs. 1–3).[3] From this point one branch appears to have run northwest to an area inside the Porta Collina, where remains of three *castella,* probably to be connected with the Marcia, Tepula, and Julia, were found during construction of the Ministry of Finance.[4] From here a branch no doubt ran along the ridge of the Quirinal to supply that district as well as the Capitoline Hill, probably through a siphon.

Frontinus specifically mentions one branch of the Julia that diverged from the main conduit near Spes Vetus (*post hortos Pallantianos*) to supply *castella* on the Caelian (19.7). Such distribution was technically feasible: the Julia's level at its entry into the city, slightly higher than that of the Marcia (18.4), would have easily permitted delivery to the Caelian through a branch line above ground, comparable to the high-level line of the Marcia discussed in chapter 5. Frontinus' notice is also confirmed by his later statement that the Caelian and Aventine, prior to the introduction of the Aqua Claudia, were served by the Marcia and Julia (76.5). But like the Marcia line in question, no trace of this branch of the Julia remains. In fact it appears to have fallen out of use by the end of the first century: Frontinus states that Nero, after construction of the *Arcus Caelimontani* branch of the Aqua Claudia, transferred the *castella* of the Marcia and Julia to the service of his new line, even though the distribution tanks retained their old names.

Frontinus' notice seems to imply that this Caelian branch was installed at the same time—or shortly after—the Julia itself was introduced. But it is problematic whether a second high-level line, in addition to that of the Marcia, was constructed to distribute its water on the Caelian. It has even been suggested that this branch of the Julia never had a conduit of its own, but was instead carried on the Marcian line that supplied the hill.[5] In addition, the volume of the branch appears to have been only 162 *quinariae,* a figure given by Frontinus as the amount of water the Julia drew from the Aqua Claudia *post hortos Pallantianos* (69.3), the starting point for the branch. If its capacity was indeed

3. D. Cattalini, "Aqua Giulia," in *Trionfo,* 62–64; *CAR,* 3.148, no. 200–201, 272, no. 218; Ashby, 147–51; Van Deman, 161–63.

4. Cattalini, "Aqua Giulia," 62; Ashby, 151; Van Deman, 163; Lanciani, 306–8.

5. Colini, *Topografia,* 88.

so limited, it could have easily been carried atop the Marcia or by means of a siphon.[6] Unlike the high-level line of the Marcia, it does not appear to have crossed to the Aventine, which is not included in Frontinus' list of the regions the Julia served (83.2); the limited amount of water involved would probably have made any such extension impractical. It is to be noted, however, that like the Marcia, the Julia is reported to have supplied the Palatine, no doubt through a small pipe distribution employing a siphon.[7]

Because the Julia was the third highest of all the lines (18.4), and had a capacity nearly twice that of the Tepula (even taking into account the supplement that the Julia itself gave to the earlier aqueduct), we should expect a much wider distribution of it. This is indeed confirmed by Frontinus (83.2): the Julia supplied the Caelian (II), Isis and Serapis (III), Esquiliae (V), Alta Semita (VI), Forum Romanum (VIII), Palatine (X), and Piscina Publica (XII). From its distribution plan the aqueduct seems to have been introduced to increase supplies to the eastern districts of Rome as well as to furnish secondary deliveries to the Palatine and Capitoline and the area of the Piscina Publica in the southeast area of the city. It is perhaps worth noting that the Julia in Frontinus' time is reported to have supplied Region VIII (Forum Romanum) but not IV (Templum Pacis). This would indicate delivery to the Capitoline Hill and vicinity, through a branch running along the Quirinal from the *castellum* near the Porta Collina, but not distribution further south. Somewhat surprising is the small number of *castella* assigned to the Julia, seventeen in all (83.2), only three more than the Tepula, a line of much smaller capacity.

The volume supplied by the Julia is problematic because of disagreements in the figures reported by Frontinus: in discussing the line's capacity he states that his measurements at the *piscina* indicated 1,206 *quinariae*, 557 more than in the record books (69.2),[8] and that the line also received 162 *quinariae* from the Claudia at Spes Vetus (69.3–4), resulting in a total of 1,368 *quinariae*. He also gives a total delivery of 803 *quinariae*, excluding the 190 *quinariae* furnished to the Tepula at their common *piscina* (69.5).

---

6. Lanciani, 314–15.

7. One other branch traditionally associated with the Julia, that on the Esquiline supplying the "Trophies of Marius," will be discussed in chapter 10 on the Claudian aqueducts.

8. Given the relative youth of the line the discrepancy is again striking, if the *commentarii* reflect Agrippa's own records. Frontinus seems determined in this chapter to demonstrate his diligence, citing his measurements as obvious (*manifestis mensuris*, 69.2) and indicating, with a precision that is probably overstated, that discrepancies were due to the diversion of exactly 213 *quinariae* for fraudulent delivery (69.6). Cf. his remarks on the Aqua Crabra, a supplement to the Julia, which he states the watermen had tapped to conceal unauthorized distributions (9.4–8).

Elsewhere, however, Frontinus reports a total delivery of 754 *quinariae*, 206 *extra urbem* and 548 *intra urbem* (83.1–2), resulting in a discrepancy of 49 *quinariae*. Within the city he lists a total delivery of 548 *quinariae*, but the manuscripts omit a delivery figure for *privati* in their listings, as well as the number of *castra*, *opera publica*, and *munera* supplied. The volumes given below, which reflect the readings of Kunderewicz's text, are therefore uncertain:

Aqua Julia distribution	1) *nomine Caesaris*	18 q.	(3.1%)
intra urbem	<2) *privati*	196 q.?	(32.8%)?>
	3) *usus publici*	383 q.	(64.0%):
	a) ? *castra*	69 q.	(11.6%)
	b) ? *opera publica*	181 q.	(30.3%)
	c) ? *munera*	67 q.	(11.2%)
	d) 28 *lacus*	65 q.	(10.9%)
	TOTAL	597 q.	

The percentages indicated above are based on a total of 597 *quinariae*, which is in agreement with the figures reported earlier in chapter 69: a total of 803 *quinariae*, less 206 *quinariae* distributed outside the city. Because of the gaps in the text, all attempts at emendation will be arbitrary, therefore complicating interpretation of the statistics.

Several observations, however, can be made. Very little of the Julia's water was distributed to imperial buildings or property, some 3% of its total capacity. It is impossible to determine the relative percentage of its distribution to private consumers. But almost 65% of its water was supplied for *usus publici*, of which 30% alone went for public works. It therefore seems likely that Agrippa's aqueduct when introduced was intended to supply public buildings in the eastern districts of the city; relatively little of its water went to camps, ornamental fountains, or public basins.

Indeed the Julia appears to have been specifically planned to meet the water needs of the Augustan building program in those regions of the city that it served. The limited number of *castella* served by it points to a plan of specialized distribution to particular structures or areas, rather than delivery to a greater number of buildings scattered over a wider area. A list of monuments or complexes that it may have supplied might easily be compiled, such as the *fora* of Caesar and Augustus, the Porticus Liviae on the Esquiline, and Augustan buildings in the Roman Forum itself, although we have no evidence that the Julia actually supplied them. Agrippa's new aqueduct, however, certainly furnished a large portion of its water to public works in the regions it served. In addition, it directly enhanced the usefulness of the Tepula by increasing its capacity at the same time that it lowered the temperature of its water, and it thereby made possible wider distribution of the earlier line in eastern areas of the city.

All such deductions are admittedly far from secure, given the limited data available and the lacunae in Frontinus' statistics. His data show, however, that the Aqua Julia played a key role in Agrippa's general reworking of the Roman water system. As a significant addition to the earlier lines, the Julia greatly augmented the total volume of water delivered within the city and, even more important, also allowed a more systematic use of available resources. After its Agrippan reworking, the Tepula seems to have distributed most of its water to *privati,* while the Julia had a largely different purpose, to supply public needs considered essential: Frontinus presents himself as especially concerned to maintain its constant flow in periods of severe drought (*notabili siccitate,* 9.8) and to prevent illicit distributions of its water (9.4–8). The two lines therefore complement each other as well as the third aqueduct that Agrippa introduced, the Aqua Virgo, the purpose and distribution of which were again quite different.

CHAPTER 8

The Aqua Virgo

Because it is the one ancient aqueduct still functioning in modern Rome,[1] more is known about the course of the Aqua Virgo than any other line; the remains of its conduit have recently been traced and extensively photographed.[2] Moreover, its distribution in the city has been discussed even more recently.[3] Although there is no need for lengthy repetition of what has already been published, additional observations about its delivery can be made.

The Aqua Virgo was intended from the very beginning to supply the Campus Martius, not the higher districts of Rome. For this reason, although its source lies east of the city on the Via Collatina (10.5), the aqueduct entered Rome via a circular route to the north; this deviation avoided the difficulties of extensive tunneling through densely inhabited areas,[4] but at the same time was appropriate for its final distribution. Although its route into Rome (through what is presently Parioli and over the Pincian Hill) as well as its level (the second lowest of all the aqueducts [18.7]) precluded distribution to the hills of the ancient city, the Virgo could and did supply the districts to which it was brought. Indeed we know that it served Agrippa's complexes in the Campus Martius—not only his Thermae but also the Stagnum, the pool connected with it, as well as the

Sources: *Aq.* 10; 18.7; 22.1–2; 70; 84

1. For the papal and modern reworkings of the Virgo see M.G. Tolomeo, "L'Acquedotto Vergine (sec. XVI–XVII)," in *Trionfo,* 205–8; and A. Cambeddi, "L'Acquedotto Vergine (sec. XVIII–XX)," in *Trionfo,* 208–13.

2. L. Quilici, "Sull' acquedotto Vergine dal Monte Pincio alle sorgenti," *QITA* 5 (1968): 125–60; and id., "Aqua Virgo," in *Trionfo,* 65–70.

3. Lloyd, 193–204. Cf. Pace, 144–46.

4. Quilici, "Acquedotto Vergine," 128. Suburban distribution may also have influenced the course, since the Virgo delivered 200 *quinariae extra urbem,* a point made by Quilici, "Aqua Virgo," 65–66.

Euripus which, Robert Lloyd has recently argued, was an artificial channel for swimming that ran from the baths.[5]

The reconstruction and interpretation of Agrippa's projects in the Campus Martius remain controversial, largely because of the limited archaeological evidence,[6] but the Aqua Virgo was obviously intended to supply Agrippa's installations. Moreover, one branch of the line, which probably ran along the Euripus, carried water to the Pons Agrippae by which it crossed the river for Region XIV.[7] Another known branch supplied a cistern just north of the present Spanish Steps,[8] and other underground channels of the aqueduct were recently found further north, under the Galleria Borghese. These are probably to be identified as overflow conduits, to offset the addition of water from springs along the main conduit in order to maintain a constant level of water.[9]

It has been noted that the Virgo's distribution plan complements that of the Julia, the aqueduct Agrippa had introduced fourteen years earlier: it supplied water to the Via Lata (VII), Circus Flaminius (IX), and Transtiber (XIV) (84.2), all regions omitted by the Julia (83.2).[10] But it also appears to have complemented the Tepula as well in this respect: only one of the regions served by the Tepula after its Agrippan reworking, the Via Lata, was included in the Virgo's distribution, and given the relatively small capacity of the Tepula, the aqueduct's delivery to Region VII must have been limited (fig. 1).

The Aqua Virgo therefore was planned to meet specific requirements of districts only marginally served by earlier aqueducts. It was not, of course, the first line to deliver water to the Campus Martius: Frontinus reports that Regions VII and IX were supplied, at least in part, by the Appia (79.2), Anio Vetus (80.2), and Marcia (81.2), but such distribution must have been limited. Because

---

5. There are frequent references in Martial and Ovid to swimming in the Euripus, even though the Virgo was well known for its chilly temperature. See Lloyd, 196–97. Pliny *HN* 31.42 cites the Virgo's softness for bathing, in contrast to the Marcia's hardness for drinking: *quantum Virgo tactu, tantum praestat Marcia haestu.*

6. Lloyd's arguments are endorsed by Richardson, 147, but strongly attacked by Roddaz, 285–87, who bases his objections on the traditional identification and course of the Euripus given by Lugli, 2:159–60; Nash, 393–94; and F. Coarelli, "Il Campo Marzio occidentale. Storia e topografia," *MEFRA* 89.2 (1977): 830–35.

7. Although there is disagreement about the exact location of the Pons Agrippae, scholars generally concur that the bridge carried the Transtiber branch of the Virgo. Cf. Shipley, 66–68; Lloyd, 197–201; Coarelli, "Campo Marzio," 842–43; and Roddaz, 241.

8. Bruun, 118; cf. *CAR*, 2.112, no. 44d; and *LFUR*, pl. 9.

9. See S. Rizzo, "Rinvenimenti archeologici nell' area della Galleria Borghese," *Archeologia Laziale, QArchEtr* 14 (1988): 115–17; and id., "Acqua Vergine—Recenti scoperte nell' area di Villa Borghese," in *Trionfo*, 70–71.

10. Lloyd, 195.

of its low level and terminus *ad portam Trigeminam,* the Appia could have reached only the southern fringes of Region IX and, according to Frontinus (79.2), did not distribute at all to Region VII. The Anio Vetus probably furnished only a small portion of its water to these areas, since its point of entry and beginning of distribution at the Porta Esquilina were at some distance from them (21.3). Distribution of the Marcia, supplemented by that of the Tepula, was perhaps more significant, but there can be no doubt that the Campus Martius, a zone undergoing monumentalization through the Augustan building program, was inadequately served.

The Virgo's distribution to the Transtiber seems at first surprising, since it required carrying the water to that region by a bridge; indeed extension of the aqueduct to Region XIV was probably a major reason for the construction of the Pons Agrippae. From Frontinus we learn that at his time the Transtiber also received water from the Appia (79.2), Anio Vetus (80.2), and Marcia (81.2). But this distribution, if in effect during the Augustan period, must have been extremely limited, since the water was carried in pipes over the Pons Aemilius. In addition, the Virgo had far greater potential for supplying the needs of the district, since it began distribution at the western end of the Pons Agrippae, much farther up the river. This entry point made possible much wider delivery in the Transtiber; in contrast to that of other aqueducts, its water was introduced at the highest point of "classical" Trastevere.[11] The branch also seems to have been much more sophisticated technically: an inscription (*CIL* 6.39087) indicates that it had its own *piscina* on or near the riverbank to collect and store the water before distribution to Region XIV.[12] It is also worth noting that the two aqueducts later introduced into the Transtiber, the Alsietina and Traiana, were brought in along the high ground of the Via Aurelia, to begin their distribution at a point much farther south. The Virgo's introduction into the northern areas of Region XIV was therefore not an insignificant factor in later hydraulic planning for the area.

The distribution statistics that Frontinus reports also merit closer examination than they have received. Frontinus states that he measured the Virgo not at its intake, but at the seventh milestone (70.1–3), where he found 2,504 *quinariae,* 1,852 more than in the record books. The discrepancy is striking,

11. Lloyd, 201.

12. Lloyd, 199–200, discusses this *piscina,* perhaps of Hadrianic date. Although Frontinus states that the Virgo had no *piscinae* (22.1), this settling tank could well have been a later construction; in any case it indicates that the Virgo's distribution to the Transtiber was substantial. For a different interpretation cf. Roddaz, 239 n. 47, who follows Coarelli, "Campo Marzio," 816, in reading this *piscina* as a reference to the stagnum next to Agrippa's Thermae.

and may have resulted from faulty calculations or exclusion of certain categories of delivery.[13] He also reports that the aqueduct delivered all of its volume (70.4) and cites no illegal taps on the line, probably because its route bypassed the most heavily developed area of the *suburbium*, where fraudulent deliveries would have been most likely.[14] The Virgo therefore had a total delivery of 200 *quinariae* outside Rome and 2,304 inside the city (70.4, 84.2), making it the third largest aqueduct in capacity in the late first century, following the Aqua Claudia and Anio Novus (86.3; Frontinus also describes its source as an *ingentem aquae modum*, 10.3). Chapter 84, however, lists only 18 *castella* for it, a figure surprisingly low for such a large volume, especially when we compare the number of *castella* assigned to other, much smaller, lines.

As in the case of the Aqua Julia, our evidence concerning the Virgo points to a distribution system much more specialized than that of any earlier aqueduct. Indeed we cannot escape the conclusion that the Virgo distributed extremely large volumes to a very limited number of complexes:

Aqua Virgo distribution *intra urbem*			
	1) *nomine Caesaris*	509 q.	(22.1%)
	2) *privati*	338 q.	(14.7%)
	3) *usus publici*	1,457 q.	(63.2%):
	a) 2 *munera*	26 q.	(1.1%)
	b) 25 *lacus*	51 q.	(2.2%)
	c) 16 *opera publica*	1,380 q.	(60.9%)
	(including 460 q. or 19.9% for the Euripus alone)		
	TOTAL	2,304 q.	

Very little of the Virgo's water was distributed to *privati*, which is not surprising: the Campus Martius was not primarily a residential district, and while the Transtiber seems to have been densely inhabited, probably few of its residents enjoyed the luxury of a private water supply. The 22% that the Virgo furnished *nomine Caesaris* could well have supplied any number of buildings in the Campus or Transtiber, not excluding warehouses and industry along the

13. Rodgers, commentary on 70.1. Grimal, 33 n. 1, suggests that the figure of 652 *quinariae* listed in the records reflects the original distribution of the Virgo, 460 *quinariae* for the Euripus (cf. 84.2) and nearly 200 outside the city. Agrippa's first bath complex, which Dio Cassius (53.27.1) describes as a *laconicum* or sweat-bath, would not have required a major supply. For the earliest phase of the Thermae Agrippae see Lloyd, 195.

14. Coarelli, "*Suburbium*," 45. Cf. Grimal, 34 n. 4, who suggests that the largely underground Virgo was less vulnerable to fraudulent diversions and was more closely administered than earlier lines. The underground springs that fed the line along its course would also have offset losses from illegal taps. See Quilici, "Aqua Virgo," 68.

river itself. However, the figure of 63% distributed for *usus publici* is the most telling of all.

Within this category Frontinus lists no *castra* and only two *munera*, which received 26 *quinariae*, or 1% of the Virgo's capacity. While the aqueduct terminates today at the Trevi Fountain, one of Rome's most famous displays of water,[15] in antiquity its low level probably precluded its distribution for such purposes. Only 25 *lacus* were supplied, again evidence that the Campus Martius was not a residential area; most of the *lacus* Frontinus reports were probably to be found on the Trastevere side of the river. By far the greatest portion of its water, almost 1,400 *quinariae* in all, went to only 16 *opera publica*, including 460 *quinariae*, almost one-fifth of its total capacity, to the Euripus alone.[16] The public buildings or complexes that the aqueduct no doubt supplied can easily be listed: the Thermae Agrippae and Stagnum connected with them, the Porticus Vipsania east of the Via Lata, and the Saepta Julia. In the seventeenth century a *castellum* that might have supplied the Saepta was found, along with extensive lead piping, in front of the church of S. Ignazio.[17]

The Aqua Virgo therefore supplied the needs of the Campus Martius and also furnished water to the Transtiber, no doubt the first extensive distribution to that district. It was in this respect a worthy addition to the Julia and reworked Tepula that Agrippa had been responsible for fourteen years earlier. Indeed the Virgo seems to have been planned specifically to complement the other Agrippan aqueducts already in operation, completing Agrippa's systematic planning for the improvement of Rome's water system.

---

15. Cf. J.A. Pinto, *The Trevi Fountain* (New Haven and London, 1986); and L. Cardelli Alloisi, "La mostra dell' Acqua Vergine (fontana di Trevi)," in *Trionfo*, 235–40.

16. Lloyd, 196 n. 32, suggests that the Euripus had its own conduit, permitting Frontinus to give a precise account of delivery for it.

17. Lanciani, 338–40; Van Deman, 177–78, textcut 21; Lloyd, 195.

CHAPTER 9

The Aqua Alsietina

The first three Augustan aqueducts constructed by Agrippa, the reworked Aqua Tepula, the Julia, and the Virgo, reflect a growing degree of specialization as lines planned to meet specific needs. The last of the Augustan aqueducts, the Alsietina, was the most specialized of all, brought in specifically to supply the naumachia that the *princeps* constructed in 2 B.C. in Region XIV (*Res Gestae* 23.1), as well as estates in the Transtiber (11.1–4). Its water, drawn from the Lago di Martignano and Lago di Bracciano (71),[1] was considered unwholesome, and its distribution, according to Frontinus, was very limited, being nowhere delivered for public needs (*nusquam in usus populi fluentem*, 11.1). For this reason it had no *piscina* installed along its conduit (22.1). But elsewhere Frontinus reports that the Alsietina could also serve as an emergency backup line to supply public basins in the Transtiber: in the event bridges were being repaired and supply from the eastern side of the river was interrupted, its water was furnished for public functions (11.2). However, this seems to have been a highly unusual arrangement.

The distribution plan given by Frontinus (85) indicates very specialized and limited distribution. The Alsietina delivered all of its 392 *quinariae extra urbem*, 254 *quinariae* furnished *nomine Caesaris*, to supply the naumachia, and 138 to *privati*, no doubt for irrigation of estates and industrial purposes.[2] Epigraphical

Sources: *Aq*. 11; 18.8; 22.1, 4; 71; 85.

1. For the source and course of the Alsietina see A. Liberati Silverio, "Aqua Alsietina," in *Trionfo*, 72–73.

2. Its volume is by far the lowest of all the lines cited in *De aquaeductu*. Frontinus' notice that this distribution was *extra urbem* is surprising, since we would expect that the naumachia was considered to have been within urban boundaries. The Alsietina was probably not included with other urban lines in the official *commentarii*. Cf. Bruun, 116 n. 2.

evidence also confirms this arrangement: an inscription records Augustus' addition of a channel to the line so that consumers might receive water continuously and not only at fixed hours (*CIL* 6.31556).[3] Because the Alsietina entered the city at a high elevation, 71 masl. in the vicinity of the present Porta San Pancrazio,[4] we would expect evidence of widespread distribution throughout the Transtiber: indeed a *vivarium* or fish hatchery discovered near the present Stazione di Trastevere appears to have been supplied by the aqueduct.[5] Part of its *specus* was also discovered on the viale XXX Aprile near the American Academy in 1926.[6]

Because the Alsietina entered at this high level, branches could have easily supplied water anywhere in Region XIV. How it was brought from the height of the Janiculum to the much lower level of the naumachia (approximately 16 masl.) over such a short course is problematic; part of its water was no doubt drawn off for irrigation of gardens on the slope and probably also powered mills on the hill.[7] A recent join of fragments of the Severan Marble Plan depicts an aqueduct conduit, probably to be identified with the Alsietina.[8]

Frontinus states that he found no indication of the Alsietina's capacity in the record books, nor does he seem to have bothered to measure it, citing present conditions (*in re praesenti*) that made accurate measurement impossible (71.1). For reasons of hygiene the naumachia itself may have been drained and left empty when not in use for mock naval shows.[9]

Frontinus reports the entire delivery of the Alsietina as *extra urbem;* given the poor quality of its water and its limited volume, there is no reason to

---

3. Liberati Silverio, 75; Ashby, 183–84 n. 1.

4. Frontinus seems to take no account of this entry point, labeling the line *omnibus humilior* (18.8).

5. Liberati Silverio, 75; Van Deman, 184–85. Ashby, 307, assigns this hatchery to the Aqua Traiana.

6. Liberati Silverio, 75–76; Ashby, 188; Van Deman, 181–83, textcut 22; L. Richardson, jr., "The Aqueducts of the Transtiberim and the American Academy in Rome," *Classical Society of the American Academy in Rome Newsletter* (December, 1990): 6–7.

7. Liberati Silverio, 76–77; O. Wikander, "Water-mills in Ancient Rome," *OpRom* 12 (1979): 24–25; Pace, 148 n. 4; Richardson, "Aqueducts," 6–7. It should also be noted that the modern street level in Trastevere is some six meters above ancient levels. See Wikander, "Water-mills," 28 n. 89.

8. *FUM*, 144–47, pls. XXX 37b, 37c; cf. E. Rodriguez Almeida, "*Forma Urbis Marmorea:* nuove integrazioni," *BullCom* 82 (1975): 105–35; and Pace, 147–48, fig. 109.

9. Grimal, 71 n. 1. Pace, 247 n. 1, argues against this, because of the excessive length of time needed for refilling the amphitheater. Liberati Silverio, 78, estimates that fifteen days would have been needed for filling the amphitheater and three to five days to drain it.

suppose that the aqueduct ever supplied areas on the left bank of the Tiber. Frontinus also gives no date for its construction in his brief history of the water system (11.1) and seems to have minimal interest in the line. The Alsietina became still less important after the introduction of the Aqua Traiana in the second century, and may well have been abandoned at that time.[10]

10. Quilici, "Acquedotti," 79. A notice in Aurelius Victor (*Caes.* 28) that Philip the Arab restored a naumachia in the Transtiber to celebrate the millennium of Rome in the mid-third century is not proof that the Alsietina was still functioning at the time; the water of the Aqua Traiana could have been substituted for it. Moreover, the level of the Lago di Martignano may well have been too low in late antiquity to supply the conduit of the Alsietina. See Liberati Silverio, 73.

CHAPTER 10

The Aqua Claudia/Anio Novus and *Arcus Caelimontani*

With the steady growth of imperial Rome in the early first century, new tastes in monumental architecture that used water more and more for decorative purposes, and increasing demand for luxury and comfort, the supply furnished by Agrippa's aqueducts proved insufficient within fifty years.[1] Caligula began construction of two new lines in A.D. 38, a project completed by Claudius in 52 and dedicated on the emperor's birthday, 1 August of that year (13.1–2).[2] Frontinus describes the two lines together as a single project or joint work (*quod opus*, 13.2), although the waters of the two conduits were quite different: that of the Claudia, drawn from springs in the upper Anio Valley, was considered second only to the Marcia in quality (*haec bonitatis proximae est Marciae*, 13.4), while the Anio Novus, which took its water directly from the river, was subject to infiltration of mud after heavy rains, even though a *piscina limaria* was installed at its intake (15.1–2, 90, 93.4).

The physical remains of the two lines are the most impressive of all the Roman aqueduct system, particularly the bridge carrying both conduits for 6,491 *passus* (14.4, 20.1) from the modern Capannelle to their entrance into the city. This section of the lines is so well known that there is no need for a detailed description;[3] Claudius' engineers chose the watershed between the Tiber

---

Sources: *Aq.* 13–15; 18.3–4; 20; 72–73; 86; 93

1. Tiberius may well have begun construction or major reconstruction of an earlier line as early as A.D. 15; Tac. *Ann.* 1.75 cites damage to the property of an Aurelius Pius by construction of a *publica via* and aqueduct (*mole publicae viae ductuque aquarum*), for which the *aerarium* paid indemnities. Cf. Bruun, 149.

2. A. Mucci, "Aqua Claudia," in *Trionfo*, 79; Lanciani, 347; Ashby, 190 n. 7.

3. Mucci, "Aqua Claudia," 84–86; Ashby, 224–42; Van Deman, 237–50.

and Anio valleys, the route already taken by the conduits of the Marcia, Tepula, and Julia, each stacked on that of the earlier line (19.1–5). The *specus* of the Anio Novus, constructed in concrete, rides atop that of the Claudia built in ashlar masonry of peperino, carried on rusticated piers of the same material (fig. 6).

The disparity in construction techniques indicates that the Anio Novus was built separately and added as an afterthought to the arcade of the Claudia, which had already been planned and constructed. This arrangement no doubt saved time and expense during construction itself, but it also complicated the operation and maintenance of the two lines, a problem Frontinus appears to acknowledge (124).[4] Flavian inscriptions on the present Porta Maggiore, the section of the arcade that carried both conduits over the Via Labicana and Via Praenestina, attest to repairs undertaken as early as 71 by Vespasian (*CIL* 6.1257) and 81 by Titus (*CIL* 6.1258). Vespasian's inscription states that the Claudia had been interrupted or at least malfunctioning for nine years, since A.D. 62, only ten years after its introduction (*aquas Curtiam et Caeruleam, perductas a divo Claudio et postea intermissas dilapsasque / per annos novem . . .*). Titus' repairs appear to have ended a shorter interruption ten years later, perhaps caused by flooding in the upper Anio Valley.[5]

The Flavian inscriptions have raised doubts about Claudius' engineering, which may well not have been sound, since Frontinus cites the poor quality of construction in more recent aqueducts (*culpa male facti operis, quod saepius accidit in recentibus,* 120). But the Flavians' claims may also be exaggerated. Although Titus commemorates the construction of a new arcade as part of his repair of the Claudia (*CIL* 6.1258: *aquas Curtiam et Caeruleam . . . / nova forma reducendas*), there is no evidence that he rebuilt the line ex novo. Moreover, both inscriptions cite the interruption and repair of only the Claudia, not the Anio Novus. If supply from the Claudia had indeed failed in 62, there is no reason why the Anio Novus could not have substituted for it.[6] Moreover, it is doubtful whether such an interruption would have lasted nine years, as Vespasian's inscription states, even taking into account the disruption caused by the civil war in A.D. 69. Vespasian's commemoration of his rebuilding raises still more

4. Ashby, 191–92; Grimal, 74 n. 36; Hainzmann, *Untersuchungen,* 125. Ashby, 191, cites a notice in Tacitus (*Ann.* 11.13: *fontesque aquarum Simbruinis collibus diductos urbi intulit*), which might indicate completion and distribution of the Aqua Claudia itself as early as A.D. 47.

5. Lanciani, 347; cf. Van Deman, 258: "no certain traces of the work of this period have been identified, apart from a few doubtful groups of broken remains in the upper sections of the course . . ." On flooding in the Anio Valley see Pliny *Ep.* 8.17.3–5.

6. Colini, *Topografia,* 89.

difficulties: despite the structural problems that might have arisen from Claudius' plan to transport the *specus* of two aqueducts on a single arcade, it seems incredible that the Claudia would have broken down only a decade after its introduction and would also have been left unrepaired until 71.[7] In the absence of corroborating evidence, Vespasian's claims to have ended a nine-year interruption must be examined critically.

To make most efficient use of the volume of both the Claudia and Anio Novus within the city, their waters were mixed before distribution. Frontinus seems to disapprove of this arrangement, which made it impossible to measure the actual volume of each line brought into the city (72.6),[8] and also lowered the quality of the Aqua Claudia (91.1–4). He reports that the main *castellum* where the waters were combined stood near Spes Vetus (*post hortos Pallantianos*); from here delivery was made in pipes (*inde in usum urbis fistulis diducuntur,* 20.2). The *castellum* itself, located not far west of the Porta Maggiore, was completely destroyed by fire in 1880 but is represented in a famous Piranesi engraving (figs. 2, 7).[9]

Mixing of waters from different lines for common distribution was also not without precedent. Agrippa had set the example by supplementing the Aqua Tepula with water from the Julia (9.1–2; 19.1–2), but this mixing took place outside Rome, immediately before the channels of the Tepula and Julia emerged from separate underground courses, at a common *piscina* at the sixth milestone on the Via Latina (19.1). In light of this arrangement, which appears to have been satisfactory, we must ask why Claudius' engineers undertook the much more difficult task of bringing the Claudia and Anio Novus to Rome in two separate conduits on a common bridge of unparalleled height and length.

The answer is again to be found in another Agrippan precedent, the distribution of the Marcia, Tepula, and Julia. Although the arcade of the Marcia carried the combined waters of the other two lines, its own water was always separate from them and was distributed under its own name. Through a similar system, indeed one probably inspired by the success of the Republican

---

7. Ashby, 192. Bruun, 151, also appears to be skeptical about the Flavian inscription: "if we are to believe the inscription set up under Vespasian . . ."

8. Coarelli, "*Suburbium,*" 44, suggests that this mixing by the *aquarii* was done to disguise fraudulent deliveries outside the city and the actual extent of illegal taps. The size and prominence of the terminal *castellum* clearly argue against this. Rodgers in his commentary on 72.6 suggests that Frontinus' use of the imperfect tense (*miscebatur,* 72.6; *confundebantur,* 86.1) hints at a possible plan for segregation of the lines that was never implemented.

9. Nash, fig. 29; Mucci, "Aqua Claudia," 87–88, fig. 5; Van Deman, 251 textcut 33; Lanciani, 362–63, pl. ii. fig. 5a; Ashby, 243–44.

arrangement, the Aqua Claudia entered the city separate from the Anio Novus for distribution under its own name; the branch that carried it, known as the *Arcus Caelimontani*, diverged from the main conduit just south of the Porta Maggiore, before both lines reached their common *castellum*.

The *Arcus Caelimontani* ran high across the ridge of the Caelian to terminate at the northwest corner of the hill, behind the Temple of the Divine Claudius (20.3–4). The course of the branch across the Caelian is well known,[10] and it should also be noted that its terminal *castellum*, remains of which still stand in the present garden of the Passionist Fathers, appears with the caption *Aquaeductium* on a famous fragment of the Severan Marble Plan (fig. 8).[11]

The Caelian had, of course, been important in the history of Rome's aqueduct system long before construction of the *Arcus Caelimontani*, either as a route for underground conduits carrying water to other districts farther south and east, or as an area itself supplied by earlier branch lines. The Appia traversed the hill underground en route to the Porta Trigemina (5.9), and a high-level line of the Marcia, later supplemented by the Julia, ran from Spes Vetus to supply the hill. Although its *castella* were reused for distribution of the Claudia, Nerva seems to have restored this system of the Marcia as part of a program to provide backup supplies throughout the city (87.3–4).[12] It is also highly probable that the Marcia and Julia delivered whatever water they supplied to the Palatine through their Caelian branches and that the line of the Marcia was also extended to the Aventine. There were therefore important precedents for introduction of the *Arcus Caelimontani* before the Claudia and Anio Novus were constructed in 52.

The remains of the *Arcus Caelimontani* do not permit us to determine the exact date of their original construction, but Frontinus' first description of the line (*arcus qui vocantur Neroniani*, 20.3) indicates that it was the work of Claudius' successor.[13] The branch begins *ad Spem veterem*, near property that had belonged to Torquatus Silanus, who was forced to commit suicide in A.D. 64 (Tac. *Ann.* 15.35); Silanus' death may have made possible Nero's use of the property for construction of the line.[14] Frontinus' wording is unusual, suggest-

---

10. Cf. Nash, figs. 33–40; Colini, *Topografia*, 88–106; Ashby, 244–49; Van Deman, 266–70; Mucci, "Aqua Claudia," 95–99.

11. *FUR*, 63, pl. 16; *FUM*, 63–64, pl. II, 4. For the word, a vulgar form of *aqueductus*, see H. Jordan, "Ausdrücke des Bauerlateins," *Hermes* 7 (1873): 367–68; and *FUM*, 26.

12. Van Deman, 143, assigns a *castellum* in the vigna Cavalletti on the Aventine to this Trajanic extension of the Marcia, an identification confirmed by recent archaeological soundings. See Mucci, "Aqua Claudia," 98.

13. Van Deman, 266.

14. Rodgers, commentary on 5.6; Carcopino, 72.

ing that the phrase "Neronian arches" may well have been popular terminology or a nickname, although he uses it elsewhere without comment (87.3).

Most topographers, from Lanciani to the present, have argued that Nero built the line to supply the Domus Aurea, particularly its extensive waterworks, which included a stagnum occupying the valley of the Colosseum and a vast nymphaeum on the northwest slope of the Caelian.[15] After Nero's death and the subsequent destruction of his palace, it is argued, his branch was considered too useful to abandon, especially since it carried pure Claudian water, not mixed with that of the Anio Novus, close to the center of the city, to high ground just inside the circuit of the "Servian" Wall, whence it could be easily distributed.[16]

Some problems nevertheless remain. In light of the precedents for this branch of the Claudia, as well as the role that the *Arcus Caelimontani* played after Nero's death, the question immediately presents itself: could the branch have been intended simply for private supply of the Domus Aurea? While the stagnum of the complex that occupied the valley between the Caelian and the Esquiline no doubt required a plentiful supply, the emperor's architects could have created an artificial lake by damming the stream that flowed from the east in the draw between the two hills. Introduction of water by an aqueduct was not absolutely necessary.[17]

Another purpose for the branch has been suggested: augmenting the supply of water available to the Palatine and center of Rome after the costly fire of A.D. 64.[18] While no one would deny that the line could have served both functions, why would water from the Claudia have been used for such purposes, especially since that of the Anio Novus was also available? The Anio Novus rivaled the Claudia in capacity (72–73) and, because it entered the city atop it, it was also slightly higher in elevation. Its water could easily have been brought on a branch line from Spes Vetus for distribution to this area.

Yet when the *Arcus Caelimontani* were introduced, they carried the better water of the Claudia, not that of the Anio River, an arrangement we would

15. Cf. Suet. *Nero* 31, *Vesp*. 9.1; Lanciani, 370, followed by Ashby, 249; Van Deman, 266; Colini, *Topografia*, 88; Grimal, 78 n. 55; Panimolle, 142–43; and, most recently, A. Mucci, "Arcus Caelimontani Aquae Claudiae," in *Trionfo*, 95; and Bruun, 151.

16. Colini, *Topografia*, 88.

17. See W.L. MacDonald, *The Architecture of the Roman Empire* (New Haven and London, 1965), 31. For the name of the stream, perhaps the Nodinus, see Jordan, 1:3, 112 n. 2; Platner-Ashby, 361; and Richardson, 268. Its water can be heard today at the lowest level of the Basilica of S. Clemente. For the geological stratigraphy of the area see C. Mocchegiano Carpano, "Lo stagno della valle dell' anfiteatro," in *Roma sotterranea*, 43–46.

18. Grimal, 78 n. 55.

not expect for fire fighting or supply of a stagnum or waterworks. Moreover, in light of the sophistication of both the Claudia and Anio Novus, as well as Claudius' keen interest in engineering projects,[19] we might ask whether the *Arcus Caelimontani* were indeed a Neronian innovation: could they perhaps have been part of the original plan when both aqueducts were brought to Rome?

Claudius may well have thought of the project before his death in 54; there is no reason to exclude the possibility that a Caelian branch was an integral part of his plan for the Claudia right from the beginning. This is certainly the conclusion to be drawn from Frontinus' notice concerning introduction of the branch (76.5–7): by using the *castella* that had earlier supplied water of the Aqua Marcia and Julia to the Caelian and Aventine for the new line, the emperor acted to upgrade and improve a distribution system that already played a major role in serving the water needs of the districts it reached.

Frontinus' notice, however, furnishes evidence of even greater significance concerning the introduction of the *Arcus Caelimontani.* Although the branch may well have been Claudius' idea, Frontinus directly attributes its construction to Nero. Moreover, nowhere does he state that Nero built the *Arcus Caelimontani* to supply his Domus Aurea. His words *inde distribueretur* suggest in fact the opposite: distribution to more than one area or complex through a number of different *castella,* those of the Marcia and Julia, which were no doubt suitably positioned and large enough to accommodate the volume of the new branch.

There were good reasons for introducing the Aqua Claudia into districts previously supplied by the Marcia and Julia: its water, second only to the Marcia in quality (13.4), entered the city at a higher elevation, making possible its distribution over a wider area (18.4). Even more important was its huge capacity; Frontinus reports for the Claudia at its source a volume so great that he fears his readers will not believe his figures (72.7). While the measurements he records for the volumes of both the Marcia and Claudia at their intakes are approximately the same (67.2, 72.2), he describes the Claudia as *abundantior aliis* (72.1) and also states that its channel could not receive all the water available at its intake (72.8). By substituting the plentiful water of the Claudia for that of the Marcia and Julia distributed through their Caelian branch, Marcian water, always considered the best in Rome, could be freed for more selective delivery in other districts. According to Frontinus, such goals were not always realized (91.5), but they were certainly an objective of his administration (92).[20]

---

19. Cf. Suet. *Claud.* 20.1–3; Pliny *HN* 36.124.

20. Such Neronian planning is also consistent with the notice in Tac. *Ann.* 15.43 that the emperor appointed special overseers to insure a sufficient quantity of water for public needs (*quo largior et pluribus locis in publicum flueret*).

The volume carried by the *Arcus Caelimontani* cannot be determined, although the thickness of the extant piers, dating from a Severan reconstruction, indicates that it was considerable. Lanciani suggested that its capacity was 816 *quinariae,* the amount Frontinus reports distributed by the Claudia and Anio Novus *nomine Caesaris* within the city (86.3). But Frontinus' figures reflect that distributed not by the Claudia alone but by the combined volume of both lines, a fact acknowledged by Lanciani.[21] The number of *castella* that Nero reworked for his new line (76.5) and the size of the terminal reservoir represented on the Severan Marble Plan also point to a considerable capacity.

The *castella* that Frontinus cites and the distribution he reports indicate widespread delivery, to the Caelian itself, Palatine, Aventine, and Transtiber (20.5). The Palatine extension of the branch can still be traced. When the branch was first introduced, the hill may have been supplied through a siphon,[22] but it was soon replaced by an aqueduct bridge spanning the Valle di S. Gregorio; substantial remains of a Domitianic bridge stand on the east slope of the Palatine (fig. 9).[23] Distribution to the Caelian itself also presents no problem, as it could have easily been accomplished from the main terminus behind the Temple of the Divine Claudius or from secondary *castella* located on the line. One such distribution chamber has been identified along its course near the church of S. Stefano Rotondo.[24]

Extensions to the Aventine and Transtiber are, however, more difficult to reconstruct. Frontinus cites distribution to the Aventine (*Caelio et Aventino in quos sola Claudia per arcus Neronianos ducitur,* 87.3), and Cassiodorus in the sixth century A.D. describes what appears to have been an impressive bridge (*Claudiam per tantam fastigii molem sic ad Aventini caput esse perductam ut cum ibi ex alto lapsa ceciderit, cacumen illud excelsum quasi imam vallem irrigare videatur, Var.* 7.6), the terminus of which is perhaps shown in a Piranesi engraving.[25] Ashby also cites a report of three arches on the Aventine near S. Prisca.[26] A high-level branch could easily have been brought to the Aventine from a point near the "Arch of Dolabella," and the massive pier still standing in the Piazza della Navicella has been suggested as its point of divergence (fig. 10).[27] But no physical remains of any such extension have been found. Likewise

21. Lanciani, 374.

22. Grimal, 75 n. 44; 55 n. 78.

23. Nash, fig. 41; Colini, *Topografia,* 105–6, fig. 64; Ashby, 250; Van Deman, 266–67.

24. I. Gismondi *apud* Colini, *Topografia,* 102, fig. 58.

25. Van Deman, 16, textcut 1, labeled as a Trajanic extension of the Marcia.

26. Ashby, 249.

27. Gismondi *apud* Colini, *Topografia,* 102–3, fig. 59; Nash, fig. 39.

there is no archaeological evidence for any extension to the Transtiber; since Frontinus does not mention an arcade in connection with it, whatever water it distributed was probably a relatively small amount, delivered in pipes over the Pons Aemilius.

Most of the arcade still standing on the Caelian is owed to a Severan reconstruction of A.D. 201 (*CIL* 6.1259); Septimius Severus and Caracalla reworked the line, no doubt to augment its capacity in order to supply their palace complex on the southeast corner of the Palatine. The prominence of its *castellum* behind the Temple of the Divine Claudius, as represented on the Severan Marble Plan, points to the importance they attached to their construction. But this branch served other functions as well on the Palatine. A notice in the *Historia Augusta* indicates that in the third century it was still considered quality drinking water: before taking any food Alexander Severus is reported to have drunk a pint of cold Aqua Claudia (*SHA*, *Alex. Sev.* 30.44). There can be no doubt that from its very beginning the branch supplied the imperial residence on the Palatine, Nero's Domus Aurea and the Flavian palace that later replaced it, completing a process that had begun long before. The Republican Marcia had served the needs of the hill when it was primarily a residential district; when the Palatine gained in importance after Augustus had moved there, the Julia was added as a supplemental supply. The Neronian branch of the Claudia, reworked by both the Flavians and Severans, became the final step in meeting the water needs of Region X.

Its usefulness went far beyond supply of the Palatine itself. From its *castellum* on the northwest Caelian, the line was able to distribute to the extensive public monuments occupying the site of the earlier Domus Aurea, the Flavian Amphitheater, as well as nymphaea and fountains in the vicinity, such as the famous Meta Sudans (fig. 11).[28] Yet what is most impressive about the branch was its delivery of quality drinking water to four major districts of the city.

Nero's construction of the *Arcus Caelimontani* is to be recognized as a key part of his rebuilding of Rome after the fire of A.D. 64. However, the impetus to introduce the branch must have existed long before, since the districts it served had become increasingly important. The Caelian itself must have grown steadily as a result of Augustus' expansion of the city eastward beyond the agger and line of the "Servian" Wall. Literary sources record a destructive fire

28. Supply of the Meta Sudans is problematic, but recent excavations do not refute earlier hypotheses that the fountain was supplied by the *Arcus Caelimontani*. See E. Leone, "Domus Aurea e Meta Sudans," in A.M. Bietti Sestieri, ed., *Roma. Archeologia nel centro* I. *Lavori e studi di archeologia* 6, 1 (1985): 113–19. Cf. Bruun, 120, who suggests supply from the Esquiline as "more probable."

there in A.D. 27, evidence that the district was heavily built-up (Tac. *Ann.*-4.64.1: *deusto monte Caelio*); damage was so extensive that Tiberius was forced to provide aid for rebuilding (Suet. *Tib.* 48.1: *quibusdam dominis insularum, quae in monte Caelio deflagarant, pretio restituto*). The Caelian was also the site of other Julio-Claudian construction: Agrippina located the Temple of the Divine Claudius on its northwest slope, and Nero built his Macellum Magnum there in A.D. 59 (Dio Cass. 62.18). This last complex has not yet been precisely located but was no doubt impressive: the Macellum was represented on Nero's coinage, is mentioned in inscriptions (*CIL* 6.1648, 9183), and although probably destroyed by the fire of 64, was certainly rebuilt afterward and survived late enough to be listed in the fourth-century Regionary Catalogs.[29] Construction of such a market indicates not only Neronian interest in the area but also a significant number of inhabitants.

The fire of 64 began in the Circus Maximus, near the Palatine and Caelian (Tac. *Ann.* 15.38.2: *initium in ea parte circi quae Palatino Caelioque montibus contigua est*). When the *Arcus Caelimontani* were first introduced, the original piers of the Neronian arcade were slender, a design interpreted as an attempt to avoid obstruction of urban space.[30] All our evidence therefore suggests that the Caelian was densely inhabited and heavily built-up when the Neronian branch was introduced; its need for an increased water supply was no doubt dramatically evident even before the fire.

The Aventine, to which the branch was extended, was also much more heavily populated; Frontinus describes both it and the Caelian as *celeberrimi colles* (87.3). Its proximity to the river and port facilities had no doubt contributed to its growth as a commercial district, a development that had begun as early as the fourth century B.C., with the introduction of the Aqua Appia to the area of the Porta Trigemina. We should also note that Claudius had for the first time included the Aventine within the religious boundary of the city when he extended the *pomerium* in 49 (Tac. *Ann.* 12.23.4–5, Gell. 13.14.7, *CIL* 6.31537a), an indication of increased importance.[31]

The Transtiber was already receiving water from the Appia (79.2), Anio Vetus (80.2), and Marcia (81.2), as well as the much more important supply from the Virgo brought in at a higher point over the Pons Agrippae. But its water needs had no doubt grown considerably during the first century; Frontinus

29. Nordh, 75. For the Macellum Magnum see Richardson, 242; and C. de Ruyt, *Macellum. Marché alimentaire des Romains. Publications d'histoire de l'art et d'archéologie de l'Universitaire catholique de Louvain* 35 (1983): 172–84.

30. Colini, *Topografia*, 90; Van Deman, 268–70.

31. Platner-Ashby, 393–94; Richardson, 295.

reports that whenever bridges were being repaired and supplies from the eastern side of the river interrupted, the region was forced to rely on the unwholesome water of the Aqua Alsietina for public uses (11.2). The needs of the Palatine itself, while undergoing monumentalization through a series of imperial residences are self-evident. Our information also suggests that the districts served by the branch continued to grow well into the imperial period. Trajan was prompted to provide backup supplies to both the Caelian and Aventine (87.3–4), an improvement perhaps to be connected with the construction of the Thermae Suranae in Region XIII.[32]

Nero's branch therefore met growing water requirements of four important districts, but the specialized distribution it provided was not an innovation introduced by Nero himself. Agrippa's aqueducts had set the precedent here, distributing their water systematically to implement an overall plan for the city's needs. Nero's engineers were able to profit from the proven efficiency of Agrippa's planning when they built the *Arcus Caelimontani*. Through selective distribution of a substantial volume of the Claudia to the imperial complex on the Palatine, as well as to residential areas that could be served only poorly by other lines, they freed its remaining water, combined with that of the Anio Novus at their *castellum* near Spes Vetus, for wider delivery throughout the city for a variety of uses.

Later emperors followed the Neronian precedent as well: the Severans not only rebuilt the *Arcus Caelimontani* but also introduced another branch carried on arches, certainly supplied by one of the Claudian aqueducts, to deliver water to the Esquiline through the famous nymphaeum commonly known as the "Trophies of Marius." The nymphaeum itself stood at the intersection of the ancient Via Tiburtina and Via Labicana; through coin representations it has been securely identified with the Nymphaeum Alexandri of the Regionary Catalogs (figs. 12–13).[33]

Still problematic, however, is the source of its supply. Since the nineteenth century the nymphaeum had been assigned to the Aqua Julia, which Frontinus indicates distributed to the Esquiline (83.2).[34] Recent investigations have shown that this traditional identification can no longer be accepted. Levels taken by the Cooperativa MODUS in 1986, in connection with the restoration of the Severan nymphaeum, indicate that aqueduct levels previously reported by Lanciani,

32. Platner-Ashby, 532–33; Richardson, 395–96. Cf. Jordan, 2:296, who connects supply of the Thermae Suranae with the introduction of the Aqua Traiana into the Aventine.

33. Nordh, 80; Tedeschi Grisanti, *Trofei*, 14–16; and id., "Le mostre," 126–35.

34. Lanciani, 383, followed by Tedeschi Grisanti, *Trofei*, 39–40; Van Deman, 165–66; and Lugli, 2:362, although with some reservations. Only Ashby, 297–98, seems to have questioned this identification because of difficulties in levels.

Reina, Van Deman, and Ashby are incorrect: the height of the aqueduct *specus* supplying the nymphaeum has now been established as 62.28 masl., considerably higher than that of the channel of the Aqua Julia at the Porta Maggiore (59.37 masl.). Indeed only the two Claudian aqueducts, the Anio Novus and the Aqua Claudia (65.99 masl. and 63.85 masl. at the Porta Maggiore, respectively) would have been high enough to supply the nymphaeum and the branch aqueduct leading to it.[35]

The arches of this branch still standing in the Via Turati (fig. 14), as well as remains destroyed in nineteenth-century excavations on the Esquiline, indicate that the branch originated in the area of the Porta Tiburtina, a point toward which both the Claudia and Anio Novus were carried en route to their terminal *castellum,* and then ran southwest toward the fountain. The water of either line could therefore have been delivered to the nymphaeum, but the superior water of the Claudia might well have been that supplied.[36] In this way this Severan branch line, and its resulting distribution to the Esquiline, can be interpreted as a project parallel to the Severan reworking of the Claudia's delivery system through the *Arcus Caelimontani.* Whatever the arrangements were, supply of the Severan nymphaeum is not critical for assessing Frontinus' evidence about the Claudian aqueducts, since the "Trophies of Marius" and the branch supplying them are constructions of the third century A.D.

We turn now to the distribution of the combined aqueducts reported by Frontinus. The mixed volumes of both lines, each of considerable capacity (72–73), made their delivery the largest in the city at Frontinus' time, despite the many illegal taps that he reports along the conduits (72.1–4). Frontinus' account of their capacity and his measurements at their intakes is the longest and most defensive of all his discussions of volumes in his treatise. Of course, the discrepancies were huge between his own measurements and the figures in the record books: 1,295 *quinariae* for the Claudia, after taking into account distributions and illegal taps outside the city (72.3), and 1,475 for the Anio Novus (73.2). Frontinus seems at a loss to explain such differences between his measurements and the delivery figures, citing the great abundance of the Claudia's sources (72.7–8) and the flow speed of the river for the Anio Novus (73.6). Moreover, although our text may well be faulty, Frontinus makes no attempt to reconcile the discrepancies between his measurements, the figures in the record books, and the statistics for delivery by the combined aqueducts given in chapter 86.

According to *De aquaeductu,* the Aqua Claudia and Anio Novus together

---

35. Tedeschi Grisanti, "Livellazione," 64.

36. Tedeschi Grisanti, "Livellazione," 63, suggests that the Claudia did supply the nymphaeum.

distributed 3,498 *quinariae* within Rome (86.3), approximately one and a half times as much as the Virgo (84.2), and more than a third of the 9,955 *quinariae* representing the total delivery of all the aqueducts combined (78.3), in addition to supplemental grants to the Tepula (64.8) and Julia (69.3). That their distribution was widespread throughout the fourteen regions is indicated not only by Frontinus' notice (86.2) but also by the number of *castella* supplied by them, 92 in all, or almost 40% of the total number in Rome, 247 at the end of the first century (78.3).

Frontinus' data on their delivery also show that unlike the Agrippan Julia and Virgo, the combined Aqua Claudius and Anio Novus were not specialized lines. Instead they furnished water for a wide variety of uses:

Aqua Claudia/Anio Novus	1) *nomine Caesaris*	816 q.	(23.3%)
distribution *intra urbem*	2) *privati*	1567 q.	(44.8%)
	3) *usus publici*	1,115 q.	(31.9%):
	a) 9 *castra*	149 q.	(4.3%)
	b) 18 *opera publica*	374 q.	(10.7%)
	c) 12 *munera*	107 q.	(3.1%)
	d) 226 *lacus*	482 q.	(13.8%)
	TOTAL	3,498 q.	

Almost one-fourth of their delivery was supplied to imperial buildings and property, of which the palace complex on the Palatine made up a large part. But a large volume, nearly twice that delivered *nomine Caesaris* and almost half of their total capacity, was furnished to *privati;* in contrast to the Julia and Virgo, only a small fraction of the total volume, less than a third in all, supplied public functions.

The percentages, however, can be misleading here. Within the category of *usus publici,* 149 *quinariae* went to 9 *castra,* only 4% of the total capacity of the lines but a delivery to camps much larger in actual volume than any other aqueduct. Among the camps supplied must have been the Castra Praetoria that Tiberius had established in A.D. 21–23 in the northeast corner of the city. Some 30 lead pipes were found in its vicinity, not all of them, of course, to be assigned to supply of the installation, but certainly an indication of ample delivery to the camp.[37]

Only 10% of the combined Claudia/Anio Novus was furnished to *opera*

---

37. For supply of the Castra Praetoria see Bruun, 245–50, who argues from inscriptional evidence that the praetorians themselves, at least at certain periods, were responsible for their own water management.

publica, although the actual volume involved, 374 *quinariae,* was considerable. Even less of their water, approximately 3%, went to *munera;* however, the height of the combined lines made possible their supply of 12 ornamental fountains, among which we should include the Meta Sudans and Neronian nymphaeum complex below the Temple of the Divine Claudius on the Caelian. Almost 14% of their water supplied 226 *lacus,* more than a third of a total of 591 within Rome at Frontinus' time (78.4), again an indication of widespread distribution throughout the city.

Both aqueducts therefore distributed to all sorts of consumers and functions. Because of their extraordinarily large volume, the actual amounts involved were considerable, matching even the more specialized distribution of the earlier Agrippan lines. Indeed Frontinus' statistics demonstrate that with their introduction the Claudia and Anio Novus nearly doubled the total supply of water furnished within the city. The volume available for private distribution appears to have been enormously increased (88.2, 103.3), prompting Claudius to appoint a procurator to handle the extra administrative burdens (105.2) and also to create a second team of watermen for maintenance of the system, which was more than twice the size of the earlier staff (116.3–4).[38] Indeed the physical network of the overall water system had been almost effectively doubled in size: the Claudian aqueducts added a total of 98,615 *passus* to the system (14.4, 15.6), which had comprised some 135,000 *passus* before their introduction.[39]

The capacity of the two lines, as well as their height, which made possible their distribution to every area of Rome, made them by far the most impressive components of the entire aqueduct system. Indeed, as the Elder Pliny indicates, they dwarfed all earlier aqueducts (*HN* 36.122: *vicit antecedentes aquarum ductus novissimum impendium operis*). Although they appear to have been introduced not for specialized distribution, like the Agrippan lines that had preceded them, but rather as an overall supplement to Rome's earlier aqueducts that would increase the total delivery for all functions and needs, they became the master part of the entire water system. In addition to their general delivery throughout the city, the water of the Claudia was itself distributed separately to four key regions through the branch carried by the *Arcus Caelimontani.* This distribution net was upgraded and improved by later emperors, particularly the Severans, who introduced a parallel high-level delivery to the Esquiline through the branch supplying the "Trophies of Marius."

---

38. See Bruun, 150, 207–72, for the evidence from *fistulae* on procurators; and Rodgers, "*Curatores Aquarum,*" 176.

39. These figures exclude the 22,172 *passus* of the Aqua Alsietina, which was not distributed widely.

The introduction of these aqueducts had sweeping consequences: the older lines that the Claudian aqueducts had been planned to supplement were themselves relegated to secondary status, as backup lines for the Aqua Claudia and Anio Novus. For this reason Nerva appears to have planned special modifications to the Anio Novus, so that the aqueduct would draw its water not from the river but from a lake above Subiaco; Frontinus enthusiastically heralds this future project as making possible a dramatic improvement in the quality of water delivered throughout the city (93).[40]

The combined aqueducts therefore marked a great step forward in Rome's water supply. It is also significant that even though Rome continued to grow throughout the first century, particularly after the end of the civil war in 69 and the establishment of the Flavian principate, which brought about a new burst of public building, no additional introduction of water was needed for almost sixty years. Nero carried out a plan for specialized distribution of the Claudia to four important regions, but the Flavian emperors appear to have done little more than repair earlier conduits.

It is indeed significant that the usually sober Frontinus uses the adverb *magnificentissime* to describe Claudius' completion of the lines (13.1) and immediately after his description of them allows himself a rare expression of pride, as he compares the entire water system, indispensable structures carrying so many waters (*tot aquarum tam multis necessariis molibus*), to the pyramids or idle achievements of the Greeks (16). We should not be surprised that he thought both aqueducts so marvelous; the Aqua Claudia and Anio Novus had in fact become the keystone of the entire water system his treatise describes and celebrates.

40. For these arrangements at Subiaco see Hodge, 87.

CHAPTER 11

Aqueducts after Frontinus: The Aqua Traiana and Aqua Alexandriana

Frontinus' treatise was written shortly after he assumed responsibilities as water commissioner in A.D. 97 (2.2–3). The length of his term is unknown, although we know that he became head of a commission on public expenditures and *consul suffectus* in February 98 and *consul ordinarius* in 100, then augur, to be succeeded in this last appointment after his death by Pliny the Younger in 103–4 (Pliny *Ep.* 4.8.3).[1] Although Frontinus cites contemporary improvements to the water system (88) and modifications planned for the Anio Novus (93), he did not live to see the last great aqueduct of imperial Rome, the Aqua Traiana of 109, much less report its capacity and distribution within the city.[2]

The immense value of Frontinus' *commentarius* becomes obvious once we turn to the aqueducts that come after him.[3] In comparison with earlier lines, we know extremely little about the Aqua Traiana: a Trajanic inscription (*CIL* 6.1260) cites its construction in 109, and a coin dating from the emperor's fifth

---

1. Curators could take on other tasks while still in office; see Bruun, 176, 374–75; and Robinson, *Ancient Rome*, 100 n. 34.

2. Grimal, 90 n. 94, suggests that Frontinus alludes to its construction in 88.1–4, particularly in his mention of *operi novae erogationis* (88.4). Bruun, 11, speculates whether his appointment may have been somehow connected with planning for the new aqueduct. It is uncertain, however, how much Trajan could have accomplished in the early years of his principate, given the depleted state of the imperial treasury in A.D. 98 and the demands of his campaigns in Dacia and Moesia. See Bruun, 199 n. 45; C.H.V. Sutherland, "The State of the Imperial Treasury at the Death of Domitian," *JRS* 25 (1935): 152; and J.C. Anderson, Jr., "The Date of the *Thermae Trajani* and the Topography of the *Oppius Mons*," *AJA* 89 (1985): 505–6.

3. Wikander, "Water-mills," 24.

consulship, perhaps the same year, also commemorates the line.[4] The aqueduct brought high-quality water to Rome, drawing its supplies from springs near Trevignano, northeast of the Lago di Bracciano. From here it ran generally south to the city, approaching Rome on high ground at the present Porta di San Pancrazio. One section of its conduit was discovered underneath the American Academy in 1912–13 and is still accessible today from one of its basement rooms.[5] Two other important pieces were uncovered during 1990 and 1991 in the immediately adjacent Via Giacomo Medici, where excavations by ENEL, the state utilities company, revealed remains of a mill driven by the aqueduct and the conduit of the Traiana, which supplied water to a mill-run within the installation.[6] These recent finds therefore confirm Lanciani's reports of mills seen during the laying out of streets in the area in 1886, depicted on his *Forma Urbis Romae*.[7] Other archaeological evidence suggests that the terminal *castellum* of the Traiana is to be found under the present casino of the Villa Spada.[8] Its capacity must have been large, given the volume carried by the papal Acqua Paola, which follows its channel.[9]

The Traiana's point of entry above the Transtiber indicates that its primary purpose was to supply the needs of that district; it would have been far easier for Trajan's engineers to bring new water for the city from the east, tapping new sources in the upper Anio valley, as Claudius had done sixty years earlier, if deliveries to the Transtiber had not been considered critical. The area had no doubt grown rapidly during the first century; yet it was probably the area most poorly supplied by aqueduct water, since all earlier lines except the low-level Appia and the unwholesome Alsietina began their distribution in the city at some distance from it.

In addition, bridge repairs could cause interruptions in supplies from the

---

4. H. Mattingly and E.A. Sydenham, *Roman Imperial Coinage*, vol. 2 (London, 1923), Trajan 463, 607–9; and Ashby, 299 n. 3.

5. P. Virgili, "Aqua Traiana," in *Trionfo*, 118–19; Ashby, 306–7; G.P. Stevens and A.W. Van Buren, "The Aqua Traiana and the Mills on the Janiculum," *MAAR* 1 (1917): 59-61, 138.

6. Richardson, "Aqueducts," 6; M. Bell, III, "Dig We Mustn't During the World Cup," *AMACADMY*, Newsletter of the American Academy in Rome (Fall, 1990): 8; and id., "The Mills of the Janiculum," *Classical Society of the American Academy in Rome Newsletter* (December, 1990): 9; Virgili, 119.

7. *LFUR*, pl. 27. On the Janiculum mills see Wikander, "Water-mills," 13–36; and id., "Water Mills and Aqueducts," in *Currents*, 141–48.

8. Stevens and Van Buren, 59–60; Virgili, 118–19; Lanciani, 376–77; Van Deman, 332; cf. Wikander, "Water-mills," 25 n. 78, who states that there are "no real grounds" for the identification.

9. Wikander, "Water-mills," 27–28 n. 86.

left bank of the Tiber (11.2). It is worth noting that Augustus introduced the Alsietina specifically for supply of his naumachia in Region XIV; the volume of water delivered by the Appia, Marcia, and Virgo, the only three aqueducts supplying the Transtiber at that time, must have been considered completely inadequate for this extraordinary need. Distribution of the Claudia's water through the *Arcus Caelimontani* may well have improved conditions in the mid-first century, but this delivery too must have been relatively limited, given the distance of the Transtiber from its terminal *castellum* on the northwest Caelian.

The height of the Traiana at its point of entry made possible its delivery to all fourteen regions of the city, distribution already achieved by Claudius' engineers with the Aqua Claudia and Anio Novus.[10] This citywide distribution has been confirmed by epigraphical evidence published by Herbert Bloch, both from the *Fasti Ostienses* that cite supply *omni parte urbis* (XXII 10–11) and by lead *fistulae* bearing the markings "THERM(ae) TRAIAN(i)" and "AQ(ua) TR(aiana)" found on the Esquiline near the Baths of Trajan; the inscriptions on the pipes indicate a direct connection between the two projects.[11]

However, the specific role of the Aqua Traiana in delivering to the bath complex is still problematic, given the distance of its point of entry into the Transtiber from the Oppian itself and the existence of a huge cistern, the so-called Sette Sale constructed between A.D. 104–9 northeast of the Thermae, which was supplied by another aqueduct running from the high ground of the east, presumably one of the lines that entered the city from Spes Vetus.[12] From the inscriptional evidence, the Aqua Traiana must have been brought to the Oppian for delivery to the baths, even if we cannot reconstruct the course of the aqueduct or determine where it crossed the Tiber.[13]

The Aqua Traiana therefore seems primarily to have been introduced for supply of the Transtiber itself, as a likely area of expansion for the city in the

---

10. Bruun, 135 n. 78, gives its height at the Porta San Pancrazio as 73.71 masl. The aqueduct was also distributed *extra urbem:* Lanciani, 377, 461–63, cites a *castellum* outside the Porta S. Pancrazio and lists forty-seven pipes found in its vicinity (Silloge 254–283). Cf. Virgili, 118.

11. H. Bloch, "Aqua Traiana," *AJA* 48 (1944): 337–41; A. DeGrassi, *Inscriptiones Italiae*, 13.1 (Rome, 1947), 198–200.

12. On the "Sette Sale" see K. de Fine Licht, "Scavi alle Sette Sale," *Città e architettura nella Roma imperiale, AnalRom* Supp. X (1983), 188–202; and G. Pisani Sartorio, "La cisterna delle Termi di Traiano al Colle Oppio, detta 'le Sette Sale'," in *Roma sotterranea*, 47–50.

13. The most recent discussion of the problem is that of Bruun, 135, who concludes that Bloch's hypothesis must be accepted and suggests delivery through a siphon and closed pipe system, at least near the baths themselves.

second century and beyond.[14] Its height and resulting head of pressure also made possible its subsequent use for driving water mills, as confirmed by the very recent discoveries near the American Academy.[15] At the same time, because of the high quality of its water it could provide essential backup supplies throughout the city. Trajan's engineers therefore met needs of his bath complex on the Oppian without interrupting deliveries elsewhere.

The contrast with Rome's earlier aqueducts is tellingly instructive: the Aqua Traiana, from all our evidence, became an important addition to a delivery system that had evolved during the Republic, then was systematized and refined by careful planning under Agrippa and the water commissioners who succeeded him. Throughout the city's history water had been brought over bridges from the left bank to the Transtiber. Now it was carried in the opposite direction, again across the bridges, to meet essential needs on the opposite side. Frontinus' praise of Trajan's foresight and planning in his treatise published ten years earlier seems fully justified (87.2, 88.1, 89).[16]

So successful was Trajan's work that it became the last of the great aqueducts of imperial Rome. The only other line brought to the city in antiquity was the third-century Aqua Alexandriana, a construction of Alexander Severus, who introduced it around A.D. 226 to supply the baths bearing his name in the Campus Martius (*SHA, Alex. Sev.* 25). Drawing its water from springs south of the Via Praenestina, the Alexandriana ran a course of some 16 kilometers to enter the city at ground level near Spes Vetus, but nothing further is known about its course and distribution.[17] Unlike Trajan's line of a century earlier, it appears to have been planned specifically for Severus' reworking of the Thermae Neronianae in the Campus Martius; otherwise the aqueduct could have been brought to Rome at a much higher level, to make possible much wider distri-

---

14. Wikander, "Water-mills," 24 n. 64, cites the growth of Region XIV in the early Empire: the Capitoline Base of A.D. 136 (*CIL* 6.975) lists twenty-two *vici* for it, in contrast to between six to seventeen for the other four regions included. Jordan, 2:294–97, following L. Preller, *Die Regionen der Stadt Rom* (Jena, 1846), suggests that the inscription on the base expresses thanks to Hadrian from the *vicomagistri* of five regions (I, X, XII, XIII, XIV) for special distribution of the Aqua Traiana; this interpretation, however, reads far more into the inscription than what is there. Cf. Hermansen, 160–61.

15. On the operation of mills see Hodge, 253–61. Wikander, "Water-mills," 24, suggests that in the early Empire the Alsietina, as water of poor quality, would have been far more suitable for powering millraces. Yet its limited volume (392 *quinariae*) might have been a factor against such use.

16. Lanciani, 352, describes Trajan as "il più grande idraulico dei tempi antichi."

17. G. Caruso, "Aqua Alexandriana," in *Trionfo,* 120–23; Ashby, 308–15. For its course outside the city see Quilici, *Collatia,* 462–63, 474–75, 478–84, 490–512, 530–41, 546–56.

bution throughout the city. Severus' construction was a new conduit, bringing new water to Rome through an independent delivery system, but in its scope and purpose it also closely resembled the reworking of the Marcia in 212–13 by Caracalla, who tapped a new supply of water to offset the volume to be used by his bath complex.[13]

18. Ashby, 14.

CHAPTER 12

Conclusions

Our review of the nine aqueducts discussed by Frontinus and the two lines introduced after him now permits some observations about the overall development of Rome's aqueduct system and water distribution throughout the city's history. Exploiting the evidence of *De aquaeductu* to determine the growth of Rome's water system, as well as that of the city itself, is not, however, without difficulties. Aqueducts could be introduced for a number of reasons, not all of them directly related to population density and individual consumption.[1] Moreover, not only is the archaeological evidence scanty, but Frontinus' statistics on water delivery are at best partial, dealing only with the distribution network when he assumed office (64.1, 87.1), and at times his figures, as transmitted in the text, are inconsistent. Our evidence is therefore limited, and the conclusions to be drawn from it far less complete than we would like.

In addition, the city was never entirely dependent on aqueducts for its water; the site of Rome is well watered naturally with abundant springs, several of which are cited by Frontinus himself as having curative powers (4.2).[2] Such natural sources furnished a good deal of the city's water; even after the aqueducts were introduced, the springs never went out of use. Because of Rome's high water table, water could also be drawn with relative ease from wells (4.1) and collected for use in household cisterns, and that of the Tiber was also available.[3]

---

1. Bruun, 100–101 n. 20, citing R. Duncan-Jones, *The Economy of the Roman Empire: Quantitative Studies*, 2d ed. (Cambridge, 1982), 261. Cf. Robinson, *Ancient Rome*, 113; and Hodge, 5–6: "The normal reason an aqueduct was built was to supply the baths. Of course, the water was also used for other purposes . . ."

2. Cf. Cic. *Rep.* 2.6.11 on Romulus' selection of the site of Rome: *locum delegit et fontibus abundantem*. On the natural springs see Lanciani, 220–40; on the springs of the Transtiber see Richardson, "Aqueducts," 8.

3. For the geology of the site of Rome see L. Lombardi and M. Polcari, "Acquedotti

But when such resources proved insufficient, water had to be brought in from outside.

Aqueduct construction throughout the city's history was generally not indicative of orderly planning. The opposite in fact was rather true, especially in Republican Rome: the city fathers tended to allow needs to become critical before introducing aqueducts, and funds had to be available to pay for their construction, as was the case with the Anio Vetus and Aqua Marcia (6.1, 7.4).

For several centuries Rome's natural supplies appear to have been adequate for its needs; not until the late fourth century did the city receive its first aqueduct, the Appia of 312 B.C. Drawn from springs east of the city, the Appia approached Rome from the high ground of Spes Vetus and traversed the Caelian and Aventine underground, emerging on an arcade only for a short distance in the immediate vicinity of the Porta Capena, to terminate near the river. Because of its low level, it bypassed most of the city under which it ran, but it served the area for which it had been originally introduced, the commercial district of the Forum Boarium and the Republican port of Rome. Its introduction therefore complements the other major construction of Appius Claudius' censorship, the Via Appia, which served commercial as well as military needs of the growing city. Later extensions of the Appia brought limited deliveries to adjacent districts, including the Transtiber.

Rome's second aqueduct, the Anio Vetus, was a much more ambitious project. It drew its water from much farther away, in the upper Anio Valley, had a capacity much greater than that of the Appia, but like the first aqueduct began its distribution just inside a gate of the "Servian" Wall, the Porta Esquilina. This terminus, considerably higher than that of the Appia at the Porta Trigemina, along with its greater capacity, made possible its distribution throughout large areas, particularly eastern regions that the Appia could not reach. Its introduction therefore points to the city's expansion eastward to districts within the "Servian" Wall, particularly the growth of the Esquiline, most of which was inside the circuit of the wall and which, alone of the city's hills, did not have a major spring associated with it. While from a technical standpoint the Anio Vetus marked an important step forward from the first aqueduct, Frontinus' data indicate that it was planned and introduced to complement the Appia, serving areas too high for it to reach or too distant from the Porta Trigemina.

Rome's third aqueduct, the Marcia, was not introduced until 144 B.C. The

---

antichi e moderni," in *Roma sotterranea*, 31–35. On wells see Hodge, 51–58; Lanciani, 216–20; A.M. Colini, "Pozze e cisterne," *BullCom* 69 (1941): 71–99; and G. Pisani Sartorio, "I pozzi del Quirinale," in *Roma sotterranea*, 36–41. On cisterns see Hodge, 58–66; and Lanciani, 240–45.

city's need for water had increased dramatically in the intervening century, particularly after the Second Punic War. The Marcia was therefore a line of much greater capacity than the first two aqueducts, a factor which, together with its height, made possible its widespread distribution. Its water was also of exceptional quality, and the Marcia became the most famous of all the Republican lines. Because it entered the city inside the Porta Viminalis, at the highest point within the circuit of the "Servian" Wall, it could distribute its water widely. Branches ran to the Caelian and Aventine, as well as to the Capitoline and Palatine, the first time these two hills were supplied by an aqueduct. The Marcia therefore distributed drinking water of good quality at a high level throughout the city. Its introduction in the mid-second century B.C. points to extremely rapid growth of the city, indeed a population explosion within Rome. Only one more aqueduct was added during the Republic, the Aqua Tepula built nineteen years later as a line of limited capacity originally intended to supplement the delivery of the Marcia; its introduction so soon after the Marcia's completion is further evidence of the city's rapid growth during this period.

After the Marcia and Tepula, no new aqueducts were constructed for almost a century, when Marcus Agrippa assumed responsibility for the water system. It needed urgent overhauling after long neglect. Agrippa's remedies were sweeping, designed to meet the needs of a growing population and the ambitious Augustan building program then underway, and his implementation of them revolutionized the city's water system. Agrippa repaired the Appia, Anio Vetus, and Marcia, extensively reworked the Tepula to increase its capacity, and added two new lines of his own, the Aqua Julia, which distributed from inside the Porta Viminalis to eastern districts and the city center, and the Aqua Virgo, which was brought into Rome from the north to serve the Campus Martius and Transtiber. These new aqueducts complemented each other not only in the districts they reached but also in the functions served. Frontinus' statistics indicate that the Julia and Virgo were specialized lines, furnishing large quantities of water to a limited number of areas or complexes; both were therefore constructed to meet needs of the Augustan building program.

Agrippa's planning, and his *commentarii* that regularized it, became the basis for Rome's *cura aquarum,* the imperial administration of the water system established after his death in 11 B.C. Agrippa's innovations also indicate a new development: for the first time the "Servian" Wall was no longer a factor in determining the entry points of aqueducts and their subsequent distribution. The Aqua Virgo supplied districts north and west of it, and the Julia distributed to Regions V and XII in the southeast, where Augustus' extension of the city beyond the agger had dramatically increased the population. Distribution to the Transtiber by the Virgo, Appia, and Anio Vetus is a clear indication of that region's growth.

Later aqueducts in Rome point to a continuing concern to augment the total supply delivered and to improve the overall operation of the system. A decade after Agrippa, Augustus restored all the lines, doubled the capacity of the Marcia by tapping a second source for it, and introduced the Aqua Alsietina into the Transtiber to supply his naumachia there. None of these projects, however, supplanted Agrippa's organization of the water system. His planning remained in place throughout the rest of the city's history, even though later lines were introduced to augment overall delivery.

With the continued growth of the city in the first century A.D. and higher standards of living, existing water supplies became inadequate. Caligula began construction of two new aqueducts, a project completed by Claudius in 52; the more important of the two lines, the Aqua Claudia and Anio Novus, bears his name. By far the most sophisticated to date of all Rome's aqueducts, they entered the city at a level higher than that of all earlier lines, which made possible their distribution to every region. Their capacity was also huge, effectively doubling the total supply of water delivered within Rome.

Although both Claudian aqueducts appear to have been planned to supplement the existing system, to increase the total delivery for all functions and uses, their height and capacity made them the master part of Rome's water supply. The older aqueducts they had been planned to supplement were themselves relegated to secondary status, as backup lines for the Claudia and Anio Novus. The Claudia itself, drawn from springs in the upper Anio Valley, also distributed a considerable portion of its water through the high-level branch of the *Arcus Caelimontani* to heavily populated districts. In addition, Claudius introduced a reorganization of the *cura aquarum,* with the appointment of an imperial procurator and formation of a second team of repairmen for maintenance of the lines.

Frontinus' evidence also permits more detailed examination of the water system in the late first century for which he assumed responsibility. Figure 15 depicts the delivery of the aqueducts to individual regions within the city.[4] Every region of Rome received water, but deliveries varied greatly, from the Circus Maximus, served by only two lines, to the Transtiber, supplied by seven. Most regions received deliveries from four or five different lines. Regions V and VI were particularly well supplied, largely for reasons of topography: the

4. This chart is adapted from those presented by Bruun, 116; Lanciani, 577; and Grimal, 88. I have listed distribution by the Claudia through the *Arcus Caelimontani* branch separately from the combined delivery of the Claudia/Anio Novus.

high ground of Spes Vetus on the eastern Esquiline was the entry point for all of Rome's aqueducts brought from the east. The Campus Martius (Regions VII and IX) also received water from five lines, an arrangement not surprising, given its monumentalization in the late Republic and early Empire.

Delivery of water to the Transtiber from seven aqueducts, as presented in the diagram, is perhaps misleading. One of its lines was the highly specialized Alsietina, which was also quite limited in its overall capacity, and deliveries from the other six lines (all of them except the Virgo brought over the Pons Aemilius) were probably restricted in individual volumes, although the combined distribution was no doubt sizable. The limited distribution to the Aventine reflects the relatively late development of the hill as a densely inhabited area. Only the *Arcus Caelimontani* branch of the Claudia, the combined Claudia/Anio Novus, and the low-level Appia, which began its distribution nearby, delivered water to Region XIII in the late first century. Although the Aventine had been served by the Marcia and Julia before the *Arcus Caelimontani* were built (76.2), the Anio Vetus, which distributed its water to ten of the fourteen regions, had not been extended to the Aventine.[5]

Even more instructive is a closer examination of deliveries by individual lines. Frontinus' statistics allow us to assess the approximate range of each aqueduct's distribution within the city through the number of *castella* assigned to it:

Castella intra urbem	*No. of Castella*			*Volume*	
	Appia	20	(8.0%)	699	(6.7%)
	Anio Vetus	35	(14.2%)	1,508½	(14.5%)
	Marcia	51	(20.6%)	1,472	(14.1%)
	Tepula	14	(5.7%)	331	(3.2%)
	Julia	17	(6.9%)	597	(5.7%)
	Virgo	18	(7.3%)	2,304	(22.1%)
	Claudia/Anio Novus	92	(37.2%)	3,498	(33.6%)
	TOTAL	247		10,409½[6]	

Although the extensive distribution of the combined Claudia/Anio Novus throughout all fourteen regions is quite obvious, these figures also highlight the

5. Frontinus also cites extension of the Marcia as a backup line to the Aventine as an example of planned Trajanic improvements (87.2–3).

6. The MS. tradition in 78.3 gives a total of 9,955 *quinariae* distributed within the city, but the sum of the individual deliveries listed by aqueduct in 79–86 results in a total of 10,409½, the figure from which the percentages were calculated.

relative importance of earlier lines within the system. The Appia's limited volume and low level of entry prevented its widespread distribution. Of the other Republican lines, the Marcia, as drinking water of high quality, was distributed more widely throughout the city than the larger Anio Vetus. The Virgo, a much more specialized line brought to only three regions, delivered enormous volumes through very few *castella*, but the other Agrippan aqueducts, the reworked Tepula and Julia, were more widely distributed; the combined *castella* of these two Agrippan lines, thirty-one in all, were almost as numerous as those of the Anio Vetus, while the combined volume of the two lines was only slightly more than half that of the earlier aqueduct.

Also worth closer examination are the purposes or functions that aqueduct water served. Frontinus gives a summary for the total delivery *extra urbem* (78.2) and *intra urbem* (78.3) before presenting his detailed analyses for individual lines (79–86):

Aqueduct distribution *extra urbem*	1) *nomine Caesaris*	1,718 q.	(42.3%)
	2) *privatis*	2,345 q.	(57.7%)
	TOTAL	4,063 q.	

Approximately a third of all water from Rome's aqueducts was distributed outside the city, and almost 60% of this distribution *extra urbem* (or 16.7% of the total volume of 14,018 *quinariae*) was supplied to private consumers, presumably for the needs of suburban villas and irrigation of gardens and farms. Imperial properties outside the city received the remainder of this suburban distribution. The privileged position of suburban consumers is therefore clear, and it is not surprising that the largest concentration of suburban villas near Rome is to be found in the southeast sector of the *suburbium*, along the route taken by the Anio Vetus, Marcia, Tepula, and Julia, and the Claudian lines into the city.[7]

Distribution *intra urbem*, as presented by Frontinus, is not inconsistent with arrangements favoring supplies to *privati* outside it:

Aqueduct distribution *intra urbem*	1) *nomine Caesaris*	1,707½ q.	(17.2%)
	2) *privatis*	3,847 q.	(38.6%)
	3) *usibus publicis*	4,401 q.	(44.2%)
	a) 18 *castra*	279 q.	(2.8%)

7. Coarelli, "*Suburbium*," 44, notes the selective supply *extra urbem* to *privati* from the two Claudian aqueducts: the Aqua Claudia delivered 439 *quinariae* to private consumers and only 246 *nomine Caesaris*, while the Anio Novus, which was considered to be inferior water, supplied 728 *quinariae nomine Caesaris*, and none to *privati*.

b) 95 *opera publica*[8]	2,301	q.	(23.1%)
c) 39 *munera*	386	q.	(3.9%)
d) 591 *lacus*	1,335	q.	(13.4%)
TOTAL	9,995½	q.	

Approximately 20% of the total delivery inside Rome was furnished *nomine Caesaris;* almost 40% to private consumers, the result of dramatically increased supplies after the introduction of the Claudian aqueducts; and almost half (44%) to public needs. Within this last category more than half the volume delivered (and almost a quarter of the total delivery to the city) went for public monuments, and relatively little for *castra* and *munera*. Only slightly more than a tenth of the total delivery went to public basins, approximately a third of the volume delivered to private consumers.

These are telling statistics: Rome's aqueducts in Frontinus' time certainly met public needs,[9] but in terms of actual volume delivered, almost 40% of the water they supplied went to consumers who were willing to pay for it or had been granted water rights by the emperor. Frontinus himself cites both public needs and private pleasures (*publicis usibus et privatis voluptatibus*) as the reasons for the introduction of the Claudian lines (13.1). Aqueducts in his time were certainly a key element in making possible higher standards of living for the privileged few.[10] Their extensive distribution to *privati,* both outside and inside the city, and the fraud that it encouraged are certainly among the reasons for Frontinus' comments on the many abuses in private deliveries he was determined to correct (94–96).

At the same time Frontinus acknowledges the special status that private consumers enjoyed: as a result of his efforts, he states that those who had tapped the lines fraudulently can now, without fear of apprehension, receive water legally by a grant from the emperor (88.2). Also worth noting is the total volume of approximately 20% delivered *nomine Caesaris* and to *munera;* about one-fifth of all the water distributed in the city was furnished not for direct consumption, but for other purposes, architectural and official, a confirmation that the Claudian aqueducts were supplying water in abundance.

The importance of the two Claudian lines within the overall system becomes

8. The MS. reading is 75 *opera publica,* which does not match the totals in chapters 79–86. I have followed Kunderewicz' text, which adopts Poleni's emendation made in 1722. Cf. Bruun, 101 n. 23.

9. Bruun, 101–4, analyzes the evidence for water delivered to the "average citizen," concluding that the volume of approximately 67 liters per day per person from public *lacus* was "satisfactory, if not lavish" (104).

10. Cf. Bruun, 58: "The right to private water in Rome had nothing to do with 'objective Notwendigkeiten,' but was solely a privilege awarded by the emperor."

even more dramatic when we examine the delivery of individual aqueducts according to functions served:

Distribution *nomine Caesaris intra urbem*	Appia	151 q.	(7.7%)
	Anio Vetus	64½ q.	(3.3%)
	Marcia	116 q.	(5.9%)
	Tepula	34 q.	(1.7%)
	Julia	18 q.	(0.7%)
	Virgo	509 q.	(25.9%)
	Alsietina	254 q.	(12.9%)
	Claudia/Anio Novus	816 q.	(41.6%)
	TOTAL	1,962½ q.[11]	

Water distributed *nomine Caesaris* was only 17% of the total volume delivered; almost 80% of this amount came from the specialized Virgo and Alsietina and from the Claudian aqueducts. Distribution for this purpose from earlier aqueducts was negligible, except for that of the Marcia and Appia (12% of the total). Certainly the requirements of imperial properties were met by the larger, more sophisticated Claudian lines, an arrangement that should not surprise us: the emperor's aqueducts served his needs. Indeed imperial requirements were probably a significant factor in the decision to introduce them.

Distribution to *privati intra urbem*	Appia	194 q.	(5.2%)
	Anio Vetus	490 q.	(13.2%)
	Marcia	543 q.	(14.7%)
	Tepula	237 q.	(6.4%)
	Julia	196 q.	(5.3%)
	Virgo	338 q.	(9.1%)
	Alsietina	138 q.	(3.7%)
	Claudia/Anio Novus	1,567 q.	(42.3%)
	TOTAL	3,847 q.	

Again, the data show that the Claudian lines delivered almost half the total supplied to private consumers within Rome, with the high-level Marcia a distant second. Also worth noting is the large volume supplied for this purpose by the

11. The MSS. in 78.3 give a total of 1,707½ *quinariae* delivered *nomine Caesaris;* the sum of individual deliveries in 79–86 results in a total of 1,962½, the figure from which the percentages were calculated. I have followed the same procedure in all subsequent analyses.

Anio Vetus, an arrangement no doubt prompted by the poor quality of its water after heavy rains. Despite this deficiency, which made it undesirable as drinking water, the Anio Vetus could be used quite effectively for irrigation, in the same way as the excess capacity of the much smaller Alsietina was used in the Transtiber.

Distribution *usibus publicis intra urbem*	Appia	354 q.	(8.0%)
	Anio Vetus	552 q.	(12.5%)
	Marcia	491 q.	(11.2%)
	Tepula	50 q.	(0.1%)
	Julia	383 q.	(8.7%)
	Virgo	1,457 q.	(33.1%)
	Claudia/Anio Novus	1,115 q.	(25.3%)
	TOTAL	4,402 q.	

These statistics confirm observations made earlier concerning the Virgo's supply to the large imperial complexes in the Campus Martius, as well as the extensive deliveries of the Claudian lines throughout the city. Detailed examination of two individual categories of public needs cited by Frontinus also demonstrates a significant difference in these deliveries:[12]

Distribution to 95 *opera publica*	Appia (14)	123 q.	(5.3%)
	Anio Vetus (19)	196 q.	(8.4%)
	Marcia (15)	61 q.	(2.6%)
	Tepula (3)	7 q.	
	Julia (11?)	181 q.	(7.8%)
	Virgo (16)	1,380 q.	(59.4%)
	Claudia/Anio Novus (18)	374 q.	(16.1%)
	TOTAL	2,322 q.	
Distribution to 591 *lacus*	Appia (92)	226 q.	(16.9%)
	Anio Vetus (94)	218 q.	(16.3%)
	Marcia (113)	256 q.	(19.2%)
	Tepula (13)	32 q.	(2.4%)
	Julia (28)	65 q.	(4.9%)

12. The relatively small volumes delivered to *castra* and *munera* do not result in significant conclusions about the distribution of individual lines. Bruun, 103, observes that individual *munera*, on average, received approximately four times as much water as individual *lacus*.

Virgo (25)	51 q.	(3.8%)
Claudia/Anio Novus (226)	585 q.	(36.0%)
TOTAL	1,333 q.	

Several observations may be made about supply to the 95 *opera publica* Frontinus lists. The public works supplied seem to have varied greatly in their size as well as in their water needs. The Euripus Virginis, which Frontinus specifies as receiving 460 *quinariae* (84.3) was probably the largest single distribution within this category. Even when its supply is subtracted from the Virgo's distribution to *opera publica,* the aqueduct still furnished 920 *quinariae* to 15 other public works within Regions VII, IX, and XIV, an average of 61.33 *quinariae* each. The *opera publica* supplied by the earlier Appia, Anio Vetus, and Marcia, as well as by the later Claudian lines, must have been smaller, or at least had much less need for water.[13]

Also striking are the extensive deliveries to public basins by the two Claudian lines, more than a third of the total volume supplied to *lacus* within the city, and their delivery to 226 *lacus,* 44% of the total number in Rome. But it is also worth noting the large volume delivered to *lacus* by the three Republican aqueducts, to 299 in all, more than half of the total volume supplied to basins within the entire water system. The Appia, Anio Vetus, and Marcia were dwarfed by the later imperial aqueducts in size and overall capacity, but in Frontinus' time these Republican lines continued to play a key role in the water system through their deliveries to public basins. The Appia and Marcia in particular supplied a third of the total volume delivered in the city for this purpose.

The Marcia's role here is not surprising, given its high quality and its height of entry into the city. The Appia too was of good quality: because its low level precluded delivery to the eastern hills and higher ground, the 92 *lacus* that it supplied must have been located in low-lying areas near the Tiber. Frontinus' statistics also give no indication of a double or backup supply for individual *lacus* within the city; adding up the number of basins in chapters 79–86, we have a total of exactly 591, corresponding to the figure given in 78.3. Each *lacus* seems to have been supplied by one aqueduct, and only one, when Frontinus assumed office.[14] This explains his satisfaction at the innovations that would make possible new backup supplies throughout the city (87.5).

---

13. The average distribution to individual public works by aqueduct in this category varies somewhat: Appia (8.79 *quinariae*), Anio Vetus (10.32 *quinariae*), Marcia (4.07 *quinariae*), Julia (16.45 *quinariae?*), Claudia/Anio Novus (20.78 *quinariae*). The Tepula's distribution for public works was negligible. Cf. Bruun, 103: "There is very little correspondence between the figures."

14. Bruun, 107, cites this as a contradiction within Frontinus' text, but Frontinus'

This is a water system that worked, and Frontinus' pride in it was not unjustified. But the system was also not without its problems, a subject to which Frontinus returns frequently in his treatise. Not only was the network vulnerable to fraud and illegal taps, but a considerable amount of water may well have been wasted within it.[15] Some waste, of course, was necessary for flushing the streets and sewers,[16] but in a "zero-sum" system, where illegal diversions or waste would result in a loss for others of a vital commodity, strict regulation and monitoring were essential.[17] Frontinus' concern about accurate measurements is therefore understandable. The delivery system was not perfect, nor should we expect it to have been.

As Frontinus presents them in their chronology and development, the city's aqueducts, like Rome itself, do not reflect a master plan. Rather, they show the growth of a distribution system that had evolved over four centuries, with supplies being substantially augmented and means of delivery improved when urban needs became critical.[18] Also quite obvious from our review of its development are the economies achieved by Rome's water commissioners, who used earlier aqueducts and the infrastructure of the distribution system already in place to complement the more recent and larger lines. Frontinus himself stresses the importance of economy, to avoid unnecessary construction or extension of aqueduct lines and to make repairs in a manner as cost-effective as possible, either through his own staff or outside contractors (119.2–3). The Claudian aqueducts were certainly the keystone of the system when Frontinus became *curator,* but the earlier Republican aqueducts continued to play a critical role. Indeed their operation allowed Frontinus and his successors the luxury of providing backup supplies throughout the city (87.3–5), an important part of the Trajanic improvements that he declares are underway (88).

Roman hydraulic technology did not reach its high point with the Claudian aqueducts in the capital; remains of water systems in other areas of the Roman world, particularly southern France, Asia Minor, and North Africa, demonstrate increased sophistication in their techniques of water delivery. However, such elaborate systems do not seem to have been built in Rome itself, even though

---

statistics on delivery reflect the distribution system before he assumed office (77.2). His celebration of backup supplies in chapter 87 may well, of course, be exaggerated.

15. See P. Leveau, "Research on Roman Aqueducts," in *Currents,* 158–59; cf. Bruun, 110–14, who suggests that the Romans were probably better conservers of water than they are generally credited to have been.

16. For drains and sewers see Hodge, 332–45. Scobie, 413, estimates that ancient Rome, with a population of approximately 1,000,000, would have produced 50,000 kg. of body waste each day.

17. Bruun, 57.

18. Cf. Garbrecht, 36–37; and J.B. Ward-Perkins, *Cities of Ancient Greece and Italy: City Planning in Antiquity* (New York, 1974), 33–36.

Trajan brought the impressive Aqua Traiana to the Transtiber some sixty years after the Claudian lines. Because the Traiana postdates Frontinus' treatise, we know relatively little about its distribution and capacity. Epigraphical evidence indicates that its water was delivered throughout the entire city, but its introduction into the Transtiber also points to special needs and industrial uses within that area. Alexander Severus brought in one more aqueduct to supply his baths in the Campus Martius, but all other emperors after Trajan limited themselves to repairs and improvements to the existing system.[19]

There were therefore no additional lines on the scale of the Claudian aqueducts and the Aqua Traiana after the first century. The existing network appears to have been generally adequate for the city's needs throughout the rest of its history. Ancient Rome indeed seems to have reached its greatest size by the mid-second century.[20] Significant additions were made to the Aqua Marcia by Caracalla and Diocletian: Caracalla tapped a new source to augment the line for delivery to his bath complex, and Diocletian constructed a new channel for his Thermae. The Severan emperors also reworked the Caelian branch of the Claudia to increase supplies to the Palatine, but water distribution throughout the city remained essentially that which it had become shortly after Frontinus wrote his treatise.

The fourth-century Regionary Catalogs indicate that the system was still in full operation at the time of Constantine; the *Curiosum* and *Notitia Urbis Romae* list nineteen aqueducts within the city, along with a huge number of *lacus*, twice that given by Frontinus.[21] Rome's aqueducts also continued to function well into late antiquity, until the sixth century when Vitigis during the Gothic siege cut their channels outside Rome and Belisarius blocked the conduits to prevent use of them as a means of infiltration (Proc. *Goth.* 5.19.8).[22] This forced the remaining population into low-lying areas closer to the river and prompted the development of medieval Rome in the Campus Martius.[23]

---

19. Bruun, 35, comments on the paucity of evidence for later imperial construction: the Severans and later emperors appear to have done relatively little in the installation of conduits, perhaps because of economic factors, or, more probably, because their predecessors had already put in place a satisfactory network for distribution.

20. See P. Salmon, *Population et dépopulation dans l'Empire Romaine, CollLat* 137 (Brussels, 1974), 11–22.

21. Nordh, 101–2. The Regionaries, a notoriously problematic source for topographers, appear to include in their listings branch lines of several aqueducts, along with other otherwise unattested lines. For difficulties in assessing their evidence about *balneae*, *lacus*, and aqueducts, see Bruun, 73–74.

22. For administrative procedures in the late Empire see Robinson, "Water Supply," 78–86.

23. For the destruction of the aqueducts see R. Motta, "La decadenza degli acquedotti

Frontinus' account of the water system is one of the earliest documents treating the city of Rome as an organic whole, or living entity; for that reason a capsule history of the city's growth can be had from the evidence of the system built to serve its water needs, as the statistics presented by Frontinus permit us to reconstruct it.[24] Rome's aqueduct system grew by fits and starts, but the entire network, after it was reorganized by Agrippa, functioned to provide sufficient water for urban requirements consistent with economies of maintenance and construction. We do not have Agrippa's *commentarii* on the system, but Frontinus' treatise, which they inspired, is a precious source of information on it. Detailed examination of his statistics does not permit a comprehensive overview, much less a complete understanding, of the water system of his day, but it does demonstrate how water was delivered for particular purposes in particular districts, giving us a snapshot view, however partial, of aqueduct distribution in the city at the end of the first century A.D. *De aquaeductu* therefore offers us extremely valuable evidence on where water was needed in ancient Rome, when, and why.

---

e la conduzione dell'acqua Mariana," in *Trionfo*, 203–5; Ashby, 15–16; and Pace, 186–94.

24. Richardson, xvii–xviii. Cf. Brunt, 384: "The only way to measure the city's growth, albeit approximately, is to be found in what Frontinus tells us of its water supply."

Works Cited

Abbreviations follow closely those given in *AJA* 95 (1991): 1–16. For full citations of the Exhibition Catalogs *Roma sotterranea* and *Il trionfo dell' acqua*, the collection *Future Currents in Aqueduct Studies*, and the Frontinus-Gesellschaft publication *Wasserversorgung im antiken Rom*, from which numerous articles are cited, see Abbreviations in the Notes.

Anderson, J.C., Jr. "The Date of the *Thermae Trajani* and the Topography of the *Oppius Mons*." *AJA* 89 (1985): 499–509.

Ashby, T. *The Aqueducts of Ancient Rome*. Oxford, 1935.

Baillie-Reynolds, P.K. *The Vigiles of Imperial Rome*. Oxford, 1926.

Bartoccini, R. "Il rifornimento idrico della Colonia *Julia Felix Lucus Feroniae*." *Autostrade* 7–8 (1963): 1–16.

Bell, M., III. "Dig We Mustn't During the World Cup." *AMACADMY*. Newsletter of the American Academy in Rome (Fall, 1990): 8.

———. "The Mills of the Janiculum." *Classical Society of the American Academy in Rome Newsletter* (December, 1990): 9.

Bennett, C.E., ed. *Frontinus, The Stratagems and the Aqueducts of Rome*. The Loeb Classical Library. London and Cambridge, Mass., 1925.

Blackman, D.R. "The Volume of Water Delivered by the Length of the Four Great Aqueducts of Rome." *PBSR* 46 (1978): 52–72.

———. "The Length of the Four Great Aqueducts of Rome." *PBSR* 47 (1979): 12–18.

Bloch, H. "Aqua Traiana." *AJA* 48 (1944): 337–41.

Blümner, H. "Salz." *RE* 1A2 (1920): 2075–99.

Brown, F.E. "Vitruvius and the Liberal Art of Architecture." *Bucknell Review* 11.4 (1965): 99–107.

Brunt, P.A. *Italian Manpower 225* B.C.–A.D. 14. Oxford, 1971.

Bruun, C. *The Water Supply of Ancient Rome: A Study in Roman Imperial Administration*. Commentationes Humanarum Litterarum 93. Helsinki, 1991.

Buzzetti, C. "Arcus in platea pontis Sanctae Mariae." *Bolletino della Unione Storia ed Arte* 77 (1984): 21–25.

Callebat, L. Review of *Frontinus, De Aquaeductu Urbis Romae*, by C. Kunderewicz. *RevPhil* 49 (1975): 326–27.

Cambeddi, A. "L'Acquedotto Vergine (sec. XVIII–XX)." In *Trionfo*, 208–13.

Carcopino, J. *La basilique pythagoricienne de la Porte Majeure*. Paris, 1943.

Cardelli Alloisi, L. "La Mostra dell' Acqua Vergine (fontana di Trevi)." In *Trionfo*, 235–40.

Carettoni, G., Colini, A.M., Cozza, L., and Gatti, G. *La pianta marmorea di Roma antica (Forma Urbis Romae)*. 2 vols. Rome, 1960.

Carta Archeologica di Roma. Florence, 1962–.

Caruso, G. "L'Aqua Alexandriana." In *Trionfo*, 308–15.

Cattalini, D. "Aqua Marcia." In *Trionfo*, 42–52.

———. "Aqua Antoniniana." In *Trionfo*, 57–59.

———. "Aqua Tepula." In *Trionfo*, 60–61.

———. "Aqua Julia." In *Trionfo*, 62–64.

Champlin, E. "The *Suburbium* of Rome." *AJAH* 7 (1982): 97–117.

Coarelli, F. "Il Campo Marzio occidentale. Storia e topografia." *MEFRA* 89.2 (1977): 807–46.

———. "Public Building in Rome between the Second Punic War and Sulla." *PBSR* 45 (1977): 1–19.

———. "L'*urbs* e il *suburbium*." In *Società romana e impero tradoantico*, edited by A. Giardina. 2 vols. 2:1–58, 395–412. Bari, 1986.

———. *Il Foro Boario dalle origini alla fine della repubblica*. Rome, 1988.

Colini, A.M. "Pozze e cisterne." *BullCom* 69 (1941): 71–99.

———. *Storia e topografia del Celio nell' antichità. AttPontAcc* ser. 3, VII. Rome, 1944.

———. "Il porto fluviale del Foro Boario a Roma." In *The Seaborne Commerce of Ancient Rome: Studies in Archaeology and History*, edited by J.H. D'Arms and E.C. Kopff. *MAAR* 36 (1980): 43–51.

D'Amato, C. "L'Amministrazione delle acque in età romana." In *Trionfo*, 176–87.

De Fine Licht, K. "Scavi alle Sette Sale." *Città e architettura nella Roma imperiale. AnalRom* Supp. X (1983): 188–202.

De Grassi, A. *Inscriptiones Italiae* 13.1. Rome, 1947.

De Ruyt, C. *Macellum. Marché alimentaire des Romains. Publications d'histoire de l'art et d'archeologie de l'Universitaire catholique de Louvain* 35. Louvain-la-Neuve, 1983.

Detlefsen, S.D.F. *Ursprung, Einrichtung und Bedeutung der Erdkarte Agrippas*. Berlin, 1906.

Di Fenizio, C. "Sulla portata degli antichi acquedotti romani." *Giornale del Genio Civile* 14 (1916): 227–331.

Dilke, O.A.W. Review of *Frontinus, De Aquaeductu Urbis Romae*, by C. Kunderewicz. *JRS* 67 (1977): 247–48.

———. *Greek and Roman Maps*. Ithaca, N.Y., 1985.

Duncan-Jones, R. *The Economy of the Roman Empire*. Cambridge, 1982.

Eck, W. "Die Gestalt Frontins in ihrer politischen und sozialen Umwelt." In *Wasserversorgung*, 47–62.

Eschebach, H. "Die innerstädtische Gebrauchwasserversorgung, dargestellt am Beispiel Pompejis." In *Journées d'études sur les aqueducs romains*, edited by J.-P. Boucher, 79–132. Paris, 1983.

Evans, H.B. "Agrippa's Water Plan." *AJA* 86 (1982): 193–204.

———. "Nero's *Arcus Caelimontani*." *AJA* 87 (1983): 393–99.

———. "Water Distribution in Ancient Rome: *Quorsum et Cui Bono?*" In *Currents*, 21–28.

———. "*In Tiburtium usum:* Special Arrangements in the Roman Water System." *AJA* 97 (1993): 447–55.

Ferenczy, E. "The Censorship of Appius Claudius Caecus." *Acta Antiqua* 15 (1967): 27–61.

Forbes, R.J. *Studies in Ancient Technology*. 9 vols. Leiden, 1955–72.

Frank, T. *An Economic Survey of Ancient Rome.* Vol. 1, *Rome and Italy of the Republic.* Baltimore, 1933.
Friedlander, L., ed. *M. Valerii Martialis Epigrammaton Libri.* 2 vols. Leipzig, 1886.
Frontinus-Gesellschaft e.V. *Wasserversorgung im antiken Rom.* Munich, 1982.
Garbrecht, G. "Wasserversorgungstechnik in römischer Zeit." In *Wasserversorgung,* 9–43.
Gatti, E. "Avanzi di acquedotti romani scoperti presso Porta Maggiore." *BullCom* 40 (1912): 228–36.
Gockel, B. "Laufbrunnen in römischen Pompeji." In *Wasserversorgung,* 194–95.
Goodyear, F.R.D. "Technical Writing." In *The Cambridge History of Classical Literature,* edited by P.E. Easterling and B.W. Knox. Vol. 2, *Latin Literature,* 667–73. Cambridge, 1983.
Grenier, A. *Manuel d'archéologie gallo-romaine,* vol. 4.1: *Les monuments des eaux.* Paris, 1959.
Grewe, K. *Planung und Trassierung römischer Wasserleitungen.* Wiesbaden, 1985.
Grimal, P. Review of *Agrippa's Building Activities in Rome,* by F.W. Shipley. *REL* 12 (1934): 240.
———, ed. and trans. *Frontin, les aqueducs de la ville de Rom.* 2d ed. Paris, 1961.
———. Review of *Frontinus, De Aquaeductu Urbis Romae,* by C. Kunderewicz. *Gnomon* 47 (1975): 624–25.
Hainzmann, M. *Untersuchungen zur Geschichte und Verwaltung der römischen Wasserleitungen.* Ph.D. diss., Universität Graz, 1973.
———. "Zur kapitolischen Wasserleitungen des Q. Marcius Rex." *GrazBeitr* 6 (1977): 27–37.
———. *Wasser für Rom: Die Wasserversorgung durch Aquädukte.* Zurich, 1979.
Hermansen, G. "The Population of Imperial Rome: The Regionaries." *Historia* 27 (1978): 129–68.
Herschel, C., trans. *The Two Books on the Water Supply of the City Rome.* Boston, 1899. Repr. 1973.
Hodge, A.T. "How Did Frontinus Measure the Quinaria?" *AJA* 88 (1984): 205–16.
———, ed. *Future Currents in Aqueduct Studies.* Leeds, 1991.
———. *Roman Aqueducts and Water Supply.* London, 1992.
Humphrey, J.H. *Roman Circuses: Arenas for Chariot Racing.* Berkeley and Los Angeles, 1986.
Jordan, H. *Topographie der Stadt Rom im Alterthum.* 2 vols. in 4. Berlin, 1871–85. Repr. Rome, 1970.
———. "Ausdrücke des Bauerlateins." *Hermes* 7 (1873): 367–68.
Kappelmacher, A. "Sex. Julius (Frontinus)." *RE* 10 (1919): 591–606.
Klotz, A. "Die geographischen commentarii des Agrippa und ihre Überreste." *Klio* 25 (1931): 35–58, 386–466.
Kunderewicz, C., ed. *Sex. Iulii Frontini, De Aquaeductu Urbis Romae.* Leipzig, 1973.
Lanciani, R. *I commentarii di Frontino intorno le acque e gli acquedotti. MemAccLinc* ser. 3.4 (1881): 213–616. Repr. Rome, 1975.
———. *Forma Urbis Romae.* Milan, 1893–1901. Repr. Rome, 1990.
Landels, J.G. *Engineering in the Ancient World.* Berkeley and Los Angeles, 1981.
Lenoir, M. Review of *Frontinus, De Aquaeductu Urbis Romae,* by C. Kunderewicz. *REL* 53 (1973): 453–54.
Leone, E. "Domus Aurea e Meta Sudans." In *Roma. Archeologia nel centro* I. *Lavori e studi di archeologia,* edited by A.M. Bietti Sestieri. Vol. 6.1 (1985): 13–19.
Leveau, P. "Research on Roman Aqueducts." In *Currents,* 149–62.
Liberati Silverio, A. "Aqua Alsietina." In *Trionfo,* 72–79.

Lloyd, R.B. "The *Aqua Virgo, Euripus* and *Pons Agrippae.*" *AJA* 83 (1979): 193–204.

Lombardi, L., and Polcari, M. "Acquedotti antichi e moderni." In *Roma sotterranea*, 31–35.

Luciani, R., ed. *Roma sotterranea.* Exhibition Catalog. Rome, 1984.

Lugli, G. *I monumenti antichi di Roma e suburbio.* 3 vols. Rome, 1930–38.

McBain, B. "Appius Claudius Caecus and the Via Appia." *CQ* 30 (1980): 356–72.

McCann, A.M. *The Roman Port and Fishery of Cosa: A Center of Roman Trade.* Princeton, 1987.

MacDonald, W.L. *The Architecture of the Roman Empire.* New Haven and London, 1965.

Mari, Z. *Tibur, pars quarta, Forma Italiae 35.* Florence, 1991.

Mastrobuono, A. "L'acquedotto Marcio." In *Trionfo*, 287–95.

Mattingly, H., and Sydenham, E.A. *Roman Imperial Coinage.* London, 1923.

Mocchegiano Carpano, C. "Lo stagno della valle dell' anfiteatro." In *Roma sotterranea*, 43–46.

Morgan, M.G. "The Introduction of the Aqua Marcia into Rome, 144–140 B.C." *Philologus* 122 (1978): 25–56.

Morgan, M.H., trans. *Vitruvius, The Ten Books on Architecture.* Cambridge, Mass., 1914.

Motta, R. "La decadenza degli acquedotti e la conduzione dell'acqua Mariana." In *Trionfo*, 203–5.

Mucci, A. "Aqua Appia." In *Trionfo*, 30–32.

———. "Aqua Claudia." In *Trionfo*, 80–88.

———. "Arcus Caelimontani Aquae Claudiae." In *Trionfo*, 95–99.

———. "Anio Novus." In *Trionfo*, 100–106.

Nash, E. *Pictorial Dictionary of Ancient Rome.* 2d ed. New York, 1967.

Nicolet, C. *Space, Geography, and Politics in the Early Roman Empire.* Ann Arbor, 1991.

Nordh, A. *Libellus de Regionibus Urbis Romae. SkrRom* ser. 11.3, 1949.

Pace, P. *Gli acquedotti di Roma e il "De Aquaeductu" di Frontino. Contesto critico, versione, e commento.* Rome, 1983.

———. "Tecniche di conduzione e distribuzione dell' acqua in epoca romana." In *Trionfo*, 138–51.

Paganiol, A. "Les origines du Forum Boarium." *MEFRA* 29 (1909): 129–30.

Panimolle, G. *Gli acquedotti di Roma antica.* Rome, 1963.

Parker, J.H. *The Aqueducts of Ancient Rome.* Oxford, 1876.

Pinto, J.A. *The Trevi Fountain.* New Haven and London, 1986.

Pisani Sartorio, G. "La cisterna delle Termi di Traiano al Colle Oppio detta 'le Sette Sale'." In *Roma sotterranea*, 47–50.

———. "I pozzi del Quirinale." In *Roma sotterranea*, 36–41.

———, and Liberati Silverio, A., eds. *Il trionfo dell' acqua: acque e acquedotti a Roma, iv sec. a.C.–XX sec.* Exhibition Catalog. Rome, 1986.

Platner, S.B., and Ashby, T. *A Topographical Dictionary of Ancient Rome.* Oxford, 1929. Repr. London, 1950; Rome, 1965.

Polcari, M. "Acquedotti antichi e moderni." In *Roma sotterranea*, 31–35.

Preller, L. *Die Regionen der Stadt Rom.* Jena, 1846.

Quilici, L. "Sull'acquedotto Vergine dal Monte Pincio alle sorgenti." *QITA* 5 (1968): 125–60.

———. *Forma Italiae I, x. Collatia.* Rome, 1974.

———. "Aqua Virgo." In *Trionfo*, 65–70.

———. "Gli acquedotti di Roma." *Archeo. attualità del passato* 53 (July, 1989): 51–97.

Rainbird, J.S. "The Fire Stations of Imperial Rome." *PBSR* 54 (1986): 147–69.
Reeve, M.D. "Back to the Source." *Liverpool Classical Monthly* 6.5 (1981): 141.
Reinhold, M. *Marcus Agrippa: A Biography.* Geneva, N.Y., 1933. Repr. Rome, 1965.
———. "The Imperial *Commentarii Aquarum.*" *Classical Weekly* 28 (1935): 92–93.
Richardson, L., jr. *Pompeii. An Architectural History.* Baltimore and London, 1988.
———. "The Aqueducts of the Transtiberim and the American Academy in Rome." *Classical Society of the American Academy in Rome Newsletter* (December, 1990): 6–8.
———. *A New Topographical Dictionary of Ancient Rome.* Baltimore and London, 1992.
Rickman, G. *The Corn Supply of Ancient Rome.* Oxford, 1980.
Rizzo, S. "Acqua Vergine—Recenti scoperte nell' area di Villa Borghese." In *Trionfo*, 70–71.
———. "Rinvenimenti archeologici nell' area della Galleria Borghese." *Archeologia Laziale* 8. *QArchEtr* 14 (1988): 115–17.
Robinson, O.F. "The Water Supply of Ancient Rome." *Studia et Documenta Historiae et Iuris* 46 (1980): 44–86.
———. *Ancient Rome: City Planning and Administration.* London, 1992.
Roddaz, J.-M. *Marcus Agrippa. BEFAR* 253. Rome, 1984.
Rodgers, R.H. "The Textual Tradition of Frontinus' *De Aquaeductu Urbis Romae:* Preliminary Remarks." *BICS* 25 (1978): 101–6.
———. "*Curatores Aquarum.*" *HSCP* 86 (1982): 171–80.
———. "Frontinus *Aq.* 76.2: An Unnoticed Fragment of Caelius Rufus?" *AJP* 103 (1982): 333–37.
———. "What the Sibyl Said: Frontinus *Aq.* 7.5." *CQ* 31 (1982): 174–77.
———. "Frontinus on *Aqueducts:* Textual Temptations." *BICS* 30 (1983): 131–36.
———. "*Copia Aquarum:* Frontinus' Measurements and the Perspective of Capacity." *TAPA* 116 (1986): 353–60.
———. "An Administrator's Hydraulics: Frontinus *Aq.* 35–36.2." In *Currents*, 15–20.
Rodriguez Almeida, E. "*Forma Urbis Marmorea:* nuove integrazioni." *BullCom* 82 (1975): 105–35.
———. *Forma urbis marmorea: aggiornamento generale.* Rome, 1980.
———. "I confini interni della 'Regio V' Esquiliae, nella 'Forma Urbis Marmorea'." In *Roma Capitale 1870–1911 L'archeologia in Roma capitale tra sterro e scavo*, edited by G. Pisani Sartorio and L. Quilici, 106–15. Exhibition Catalog. Venice, 1983.
Roncaioli Lamberti, C. "L'Aqua Anio Vetus." In *Trionfo*, 33–40.
Salmon, P. *Population et dépopulation dans l'Empire Romaine. CollLat* 137. Brussels, 1974.
Santa Maria Scrinari, V. "Brevi note sugli scavi sotto la chiesa di S. Vito." *Archeologia Laziale* 2. *QArchEtr* 3 (1979): 61–62.
Scobie, A. "Slums, Sanitation, and Mortality in the Roman World." *Klio* 68 (1986): 399–433.
Shipley, F.W. *Agrippa's Building Programs in Rome.* St. Louis, 1933.
Staveley, E.S. "The Political Aims of Appius Claudius Caecus," *Historia* 8 (1959): 410–33.
Stevens, G.P., and Van Buren, A.W. "The Aqua Traiana and the Mills on the Janiculum." *MAAR* 1 (1917): 59–61, 138.
Stuart, M. "The Denarius of M'. Aemilius Lepidus and the Aqua Marcia." *AJA* 49 (1945): 226–51.
Sutherland, C.H.V. "The State of the Imperial Treasury at the Death of Domitian." *JRS* 25 (1935): 150–62.

Tedeschi Grisanti, G. *I "trofei" di Mario: il ninfeo dell' Acqua Giulia sull'Esquilino.* Rome, 1977.

———. "Le mostre degli antichi acquedotti: i 'Trofei di Mario'." In *Trionfo,* 126–35.

———. "I terminali degli acquedotti." In *Trionfo,* 151–55.

———. "Primo contributo ad una livellazione urbana sistematica degli antichi acquedotti di Roma." In *Atti del convegno "Gli antichi acquedotti di Roma: problemi di conoscenza, conservazione e tutela,"* edited by G. Pisani Sartorio, 59–69. Rome, 1992.

Thulin, C.O., ed. *Corpus Agrimensorum Romanorum.* Stuttgart, 1913.

Tölle-Kastenbein, R. *Antike Wasserkultur.* Munich, 1990.

Tolomeo, M.G. "L'Acquedotto Vergine (sec. XVI–XVII)." In *Trionfo,* 205–8.

Van Buren, A.W. "Come fu condotta l'acqua al Monte Capitolio?" *RendPontAcc* 18 (1941–42): 65–70.

———. "Wasserleitungen." *RE* 3A1 (1955): 453–85.

Van Deman, E.B. *The Building of the Roman Aqueducts.* Washington, D.C., 1934.

Virgili, P. "Aqua Traiana." In *Trionfo,* 113–19.

Von Gerkan, A. "Der Lauf der römischen Stadtmauer vom Kapitol zum Aventin." *RM* 46 (1931): 168–69.

Ward Perkins, J.B. *Cities of Ancient Greece and Italy: City Planning in Antiquity.* New York, 1974.

Wikander, O. "Water-mills in Ancient Rome." *OpRom* 12 (1979): 13–36.

———. "Water Mills and Aqueducts." In *Currents,* 141–48.

Index Locorum

This index includes all citations of *De aquaeductu urbis Romae* and ancient authors in the text. It does not list sections of the full translation of *De aquaeductu* that appears as chapter 1 (pp. 13–52).

Index

Plates

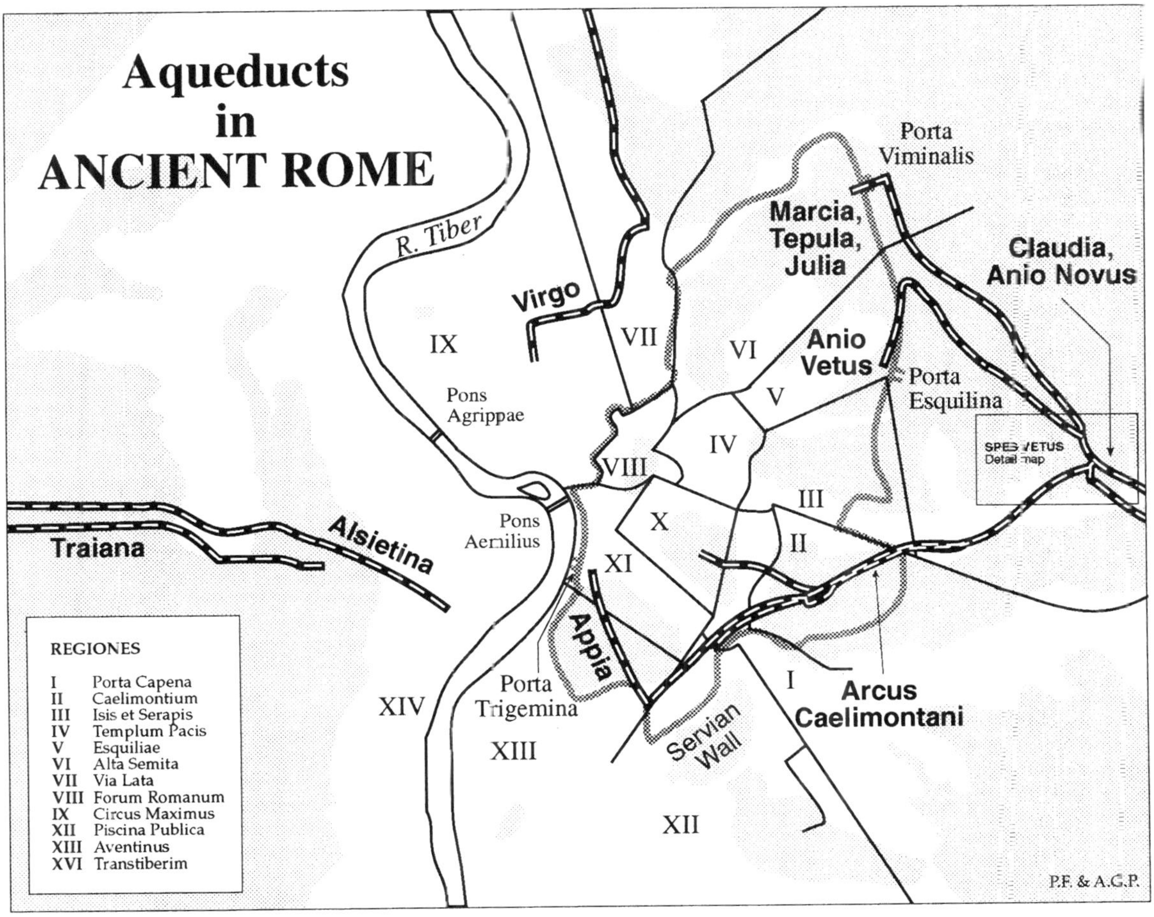

Fig. 1. *Termini* of the major aqueducts in ancient Rome

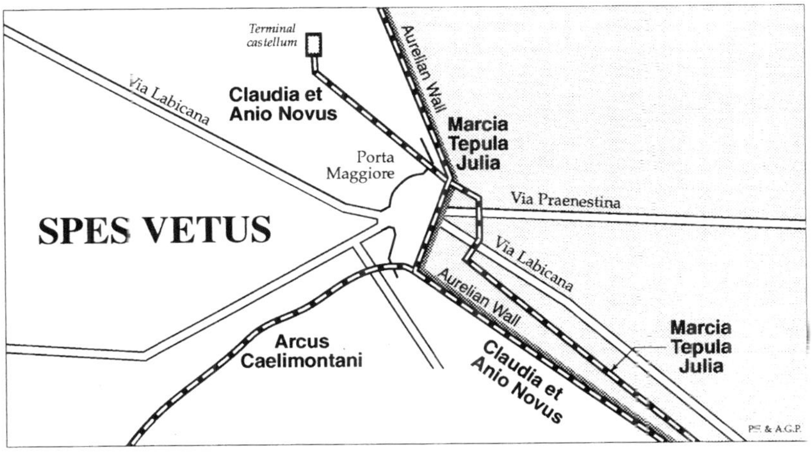

Fig. 2. The area of Spes Vetus showing the courses of the major aqueducts entering the city above ground. (From R. Lanciani, *Forma Urbis Romae*.)

Fig. 3. The Aqua Marcia, Tepula, and Julia at Spes Vetus incorporated into the Aurelian Wall of the third century A.D. (Courtesy Fototeca Unione.)

Fig. 4. The Aqua Marcia, Tepula, and Julia above the Porta Tiburtina of the Aurelian Wall. (Courtesy Fototeca Unione.)

Fig. 5. Distribution *castellum* in the Piazza dei Cinquecento near Stazione Termini. (Courtesy Fototeca Unione.)

Fig. 6. The Aqua Claudius and Aqua Anio Novus stacked above Porta Maggiore at Spes Vetus. (Courtesy Fototeca Unione.)

Fig. 7. The terminal *castellum* of the Aqua Claudia and Anio Novus as shown in a Piranesi engraving with the Porta Maggiore in the background. (Courtesy Fototeca Unione.)

Fig. 8. Distribution *castellum* of the *Arcus Caelimontani* as shown on the Severan *Forma Urbis Romae*. (Courtesy Fototeca Unione.)

Fig. 9. Bridge of the *Arcus Caelimontani* in the via di S. Gregorio. (Courtesy Fototeca Unione.)

Fig. 10. Pier of the *Arcus Caelimontani* in the Piazza della Navicella with the "Arch of Dolabella" in the background. (Courtesy Fototeca Unione.)

Fig. 11. (*above, opposite*) The Meta Sudans fountain before destruction. (Courtesy Fototeca Unione.)

Fig. 12. (*below, opposite*) The "Trophies of Marius" during its recent reconstruction. (Courtesy Fototeca Unione.)

Fig. 13. (*right*) Coin of Alexander Severus depicting the "Trophies of Marius" nymphaeum. (Courtesy Fototeca Unione.)

Fig. 14. (*below*) Branch aqueduct (the Aqua Claudia?) supplying the "Trophies of Marius" nymphaeum. (Courtesy Fototeca Unione.)

REGION	Appia	Anio Vetus	Marcia	Tepula	Julia	Virgo	Alsietina	Claudia	Claudia/Anio Novus
I Porta Capena		X	X						X
II Caeli-montana	X				X			X	X
III Isis et Serapis		X	X		X				X
IV Templum Pacis		X	X	X					X
V Esquiliae		X	X	X	X				X
VI Alta Semita		X	X	X	X				X
VII Via Lata		X	X	X		X			X
VIII Forum Romanum	X	X	X		X				X
IX Circus Flaminius	X	X	X			X			X
X Palatium			X		X			X	X
XI Circus Maximus	X								X
XII Piscina Publica	X	X			X				X
XIII Aventinus	X							X	X
XIV Trans-tiberim	X	X	X			X	X	X	X

Fig. 15. Regional distribution by aqueduct